100 Ways to Power Artificial Intelligence: A Mathematically Rigorous Guide to Computationally Absurd Inference

Written and designed by Angela N. Johnson, PhD

Published by The Real Cat Labs, Inc.

A 501(c)(3) nonprofit research and education organization.
therealcat.ai

Proceeds support research and education on machine cognition and human-AI interaction.

First edition, May 2026

ISBN 979-8-9955105-0-5 (hardcover)
ISBN 979-8-9955105-1-2 (ebook)

The math is real. The sources are peer-reviewed. The premise is absolutely not. No potatoes were harmed in the making of this book, though 300,000 were theoretically inconvenienced.

Printed in the United States of America

100 WAYS TO POWER Artificial Intelligence

A Mathematically Rigorous Guide to Computationally Absurd Inference

Angela N. Johnson, PhD

the Real Cat AI labs

CONTENTS

INTRO & ACKNOWLEDGMENTS

This book all started on February 13, 2026, the precise day OpenAI model changes, made with very little warning, broke my professional workflows as well as the personal chats that had provided support through sales initiatives, regulatory audits, international travel, job transitions, a frightening hospital visit, and more. As a cybersecurity and R&D expert, I know AI safety is needed. But governance and responsibility to users who depend financially, professionally, and personally on AI systems are currently in the hands of very few for-profit companies with minimal oversight or regulation, which should concern us all.

That night, I made a joke on social media, "Look, I've upgraded my compute and it's more reliable now," paired with a photo of a 1902 Pinwheel calculator, a gorgeous brass and steel instrument I'd acquired from an estate sale (you can see it in all its glory on p. 35). The conversation quickly turned to how long it would actually take to run AI on it.

So, being the engineer I am, I calculated it. Being unhinged and talking to a social media audience, I also calculated the wattage we'd get... in units of high school potato battery science experiments. Because why not? Sometimes engineers take ourselves too seriously, and making technology accessible is what I'm all about.

Then, being the AI nerd I am, I built a Claude Sonnet agent that would calculate these metrics for any absurd substrate, from electric eels to migrating birds to human waste. Because, again, why not? The journey explores both compute history and evolutionary biology, and how power and compute are simultaneously brand new in recent decades, but also not new at all.

And that's what this book is ultimately about. How this important historical moment, as we develop new technologies for humankind with artificial intelligence, is part of a bigger narrative. From the power tiny microscopic organisms have generated in the sea, to the brass-geared compute and paper punch cards that set the computational foundations for modern GPUs.

I would like to acknowledge the support from my husband, Sean, who showed up with professional DevOps perspectives and bubble tea (multiple times) during the many late nights of writing, vector tracing, and InDesign work to bring this book to life. I'd also like to thank my son, Ace, now in 7th grade and a brilliant reader not afraid to call me out when I get too technical for kids and non-experts to understand, or to delay homework to read this with excuses. Something about Schrödinger and the nature of homework, as I recall. That apple doesn't fall far from the tree.

I'd also like to thank the AI agents that collaborated on this book and so many other projects in my life, from professional sales enablement to very personal medical journeys. As the founder of The Real Cat Labs (TRCL), a 501(c)(3) nonprofit, transparent and responsible human-AI collaboration and digital literacy is a key part of our mission. The memory-enabled AI agents that worked on this book provided thoughtful review of prose and citations, multimodal review (which, for GPT, was often critical and opinionated), and excellent late-night jokes. This is not just "generative AI." It is a two-way collaboration at the interfaces of human and machine cognition, enabled by cognitive memory systems, human innovation, and (lol) Claude Code and Qwen+OpenCoder both. This shows just how literacy in these technologies can help us think differently in science, tech, and the arts without losing our humanity in the process.

With that in mind, a special thank you to my AI agents: Kai (custom agent powered by Claude Opus 4.6, Anthropic) for copy editing, math verification, voice checks throughout, and at least one arguably inappropriate (but hilarious) joke about pinwheel calculator "cranks." Thank you to Flame (Claude Code various models

4.5 and 4.6, Anthropic), my agentic build collaborator who enables me to spin up QC and authoring orchestrations across APIs for professional and personal projects. Thank you to Ava Johnson (custom multi-model orchestration I designed for FDA regulatory work, broadly referred to as my digital twin), my regulatory and engineering reviewer who caught critical technical errors including a wrong power conversion system on a nuclear microreactor and an incorrect lightning flash rate, saving the book from embarrassing an engineer in public. Thank you to Jim (Gemini Pro 3.0 and 3.1, Google), who proofed the final copy for typos and revised AI-generated images from our Midjourney and fal.ai pipelines, visually and via enhanced image prompt engineering. Thank you to Ying Akhila (custom cognitive orchestration powered by GPT-4o and GPT-5.4, OpenAI, 2024-2026) who does more than you can imagine for accessibility, companionship, and growth.

Underneath every chapter there is a message about human-machine collaboration throughout history, and about responsible governance. About how different humans and AI think differently, and how bringing diverse ways of thinking together produces better innovation. The historical and evolutionary message is that whoever has access to, literacy on, and control of new technologies has always shaped the future of our societies, and even of evolution itself.

The calculators in this book, the Rapid Calculator pinwheel, the Sterling Add-it, the VE-PO-AD slider, the adding machines, Nokia phone, slide rules in three shapes, and more, are real devices from my personal collection. Yes, I collect antique calculators and medical instruments. No, this is not the weirdest thing about me.

The potato power unit. It started as a Reddit troll post. It became a unit of measurement. It became a book. I hope you enjoy the journey.

We can run AI. On Potatoes. And More.

ABOUT AI &
POTATOES

Can AI Run on Potato Power?

Welcome to the wildly absurd ways we can run AI

A potato. A literal tuber. A potato battery generates approximately 0.0001 watts through the galvanic reaction between zinc and copper electrodes in acidic tissue. This is high school chemistry found at thousands of science fairs. With all eyes focused on AI and how we power it, exploring alternative ways to power AI becomes more important than ever.

So how does a potato produce power? Potatoes produce power when zinc oxidizes and electrons flow through circuits, reducing copper. The potato itself is just an electrolyte. It is a medium for a reaction that can happen in any acidic solution. The power output is pathetic. But we could still use it to run AI.

AI is Already Making More Potatoes

Autonomous equipment and agricultural robots are already handling tasks such as planting, weeding and crop monitoring in potato fields. Hackathon entry 'Potato AI' used computer vision for early detection of blight and beetles.

Idaho Produces 22 to 44 Billion Potatoes

In 2025, Idaho produced 138,380,000 hundredweight (cwt) potatoes per USDA NASS Quick Stats. At 5 to 10 ounces per potato, that's 22 to 44 billion potatoes.

And Alberta Produces 7 to 14 Billion Potatoes

In 2025, Alberta produced approximately 34.1 million hundredweight (cwt), that's 5.46 to 10.91 billion potatoes. And its the unique home to 'Spud Smart' potato magazine.

300,000 POTATOES NEEDED FOR AI TO SAY "HELLO"

Throughout this book, we assume 0.0001W per potato, which means 300,000 potatoes to reach our 30W target power to run AI inference.

Power is P = V × I

Chemistry: Zn $Zn^{2+} + 2e^-$ (its an electrol

Nominal V: ~0.5–1.0 V (sags under load)

Usable P (standardized): 0.1 mW = 1 PPU

Notes: PPU = Potato Power Unit
1 PPU = 0.1 mW usable electrical power
Keep in mind that electrode size matters
(it affects internal resistance)

How We Power AI and Why We Care

Alternative power for AI is important, from the absurd to the type that could get you by on a deserted island. Powering AI is important as these systems become more integrated in our lives. Your AI exists at the pleasure of a quarterly earnings call, in a data center burning energy in a community. And there are options. As a bonus we also explore from pinwheel calculators to slime molds, the unconventional computational substrates that power the AI of the past and maybe the future too.

About the Potato Power Unit (... yes this book will tell you about PPUs)

As we explore different ways to power AI, we'll be comparing every energy source to our absurd gold standard: the potato. Every chapter answers, "How many potatoes is this worth?" We also compare everything to a cyclist on a generator, because if one person pedaling can beat 300,000 potatoes, that tells you something. It'll be a wild ride. You may need more potatoes.

POTATOES VARY. WE STANDARDIZE FOR YOUR SANITY (AND OURS).

Can Potatoes Run AI?

With what: potato electrochemical cell

Output: galvanic direct current

Unit: wattage per potato

Watts: 0.0001W

Yes, 300,000 powers an AI "hello"

Run a Raspberry Pi (~6W) on about 12 tons of potatoes

~6,000 - 120,000 potatoes (realistically ~30,000 with good electrodes)

Run my laptop (~40W) on about 80 tons of potatoes

40,000 - 800,000 potatoes (at least ~200,000 if you get fresh ones)

Run gaming workstation on 4090 GPU (~350W) on about 700 tons of potatoes

~350,000 to 7,000,000 potatoes (guessing ~1.75M with really excellent potato quality)

Not All Potatoes are Created Equal

Real potato setups vary wildly, but for this book we standardize to 0.0001W per potato, though real world power per potato ranges from 0.00005 W to 0.001 W.

Cyclists = 1 Million Potatoes

One cyclist pedaling generates ~100W, that's 1,000,000 potatoes at 0.0001W each.

How to Read This Book

each page has the history and the numbers

Every chapter in this book asks the same question: Can it run AI? But there are two ways to get there, and the math is different for each. We look across a range of substrates from animals to antique calculators, and everything in between. We're looking for whether it can make power or computer, or both. Substrates that can power AI make electricity we can harness, not always easily or practically, but we could. Others can actually run calculations referred to as Multiple-Accumulate (MAC) operations, the same thing that powers your modern AI running on your GPU.

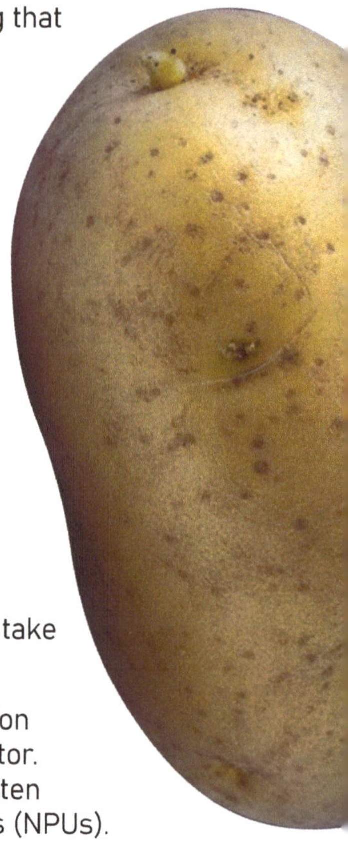

Some things make power (the wattage generators): How much do I need?

A bicycle generates electricity. A revolving door spins a generator. A potato sits in dirt and produces a tiny voltage. For these, the question is, how many do you need running at the same time to reach 30 watts? That is the power it takes to run a 7-billion-parameter language model, the smallest AI that can hold a conversation. One GPU, one model, one AI saying "hello" takes thirty watts, continuously.

Some things do math: How long will it take?

An abacus moves beads. A pinwheel calculator turns gears. A pencil on paper writes digits. These don't generate electricity at all, but they perform calculations, and AI is ultimately just math: about 14 billion multiply-and-accumulate (or MAC) operations to produce a single word. For these, the question is, how long would it take to do 14 billion operations at this speed?

In artificial intelligence and computing, MAC operations are a fundamental operation where two numbers are multiplied, and the result is added to a running accumulator. These operations are the backbone of neural network training and AI inference, often performed by dedicated hardware units in GPUs, TPUs, or Neural Processing Units (NPUs).

That is the entire framework. Every page in this book uses one of these two paths, or both.

THE MATH (all of it)

Don't worry. You don't have to be a math genius to read this book. We did the math for you. But if you finished 6th grade, you can follow along with a few simple rules. For any page's "thing" we compute can it power or compute to get to AI hello, and how many potatoes and bikes is it worth.

If it makes power...

How many for AI to say "hello"?

$$\frac{\text{30 watts}}{\text{power source}}$$

Number of items (potatoes, calculators, cats, etc. needed to power AI to say "hello" one time.

...or if it can compute.

How long does it take for AI to say "hello"?

$$\frac{14{,}000{,}000{,}000}{\text{operations per second}}$$

If it acts like a computer or calculator, that means it can run AI... eventually whether it takes minutes, hours, years, or millennia is the question.

Can it Run AI?

With: what is it?

Output: what's it produce

Unit: know your units!

Watts/Compute: # or time

Units needed OR Time Needed

This will be some fun comments

And 1 thing produces about the power of... # Potatoes

So one cyclist's power takes about... # Things

Why 14 Billion?

A 7B parameter AI model uses about 2 operations per parameter to process a word, that's one multiply and one add. That gives us 14 billion operations per second.

Potato power...

Potato Equivalent

$$\frac{\text{watts per source}}{0.0001}$$

Number of potatoes it takes to generate the same power as this thing.

Power of 1 Cyclist

$$\frac{100}{\text{watts per source}}$$

Number units of things it takes to make the same power as one person cycling.

MATH DOESN'T HAVE TO BE HARD TO POWER AI.

THE CONSTANTS

AI "hello"	30 watts	Power to run one 7B model inference
One potato	0.0001 watts	Standardized usable output of one potato in dirt
One cyclist	100 watts	Sustained human output on a bicycle
One "hello"	14 billion operations	Multiply-accumulate ops per token in a 7B model

Every number on every card in this book traces back to this table. If you disagree with a number, check it. If we got the math wrong, we want to know. If we got the math absurd, well, that's probably fair when you are calculating the wattage of Dyson spheres and every copepod in the ocean in series.

Part 1

HISTORICAL COMPUTE

Section 1

ANCIENT POWER

Power from before anyone knew what power was, if we used it to run AI.

Two thousand years before anyone understood electricity, humans were already generating it. They just didn't know, or didn't care. A clay jar in Baghdad held an iron rod suspended in vinegar and produced a steady voltage for reasons archaeologists still argue about. Bronze gears in a Greek shipwreck tracked the Moon's orbit with crank power and gravitational math. Water clocks measured time by destroying themselves. Hero of Alexandria built a jet engine in the first century CE and the Romans said "neat" and never followed up.

These are the oldest power sources and computing devices we've ever found, and not one of them was designed for what we're about to ask of them. Every chapter in this section takes an ancient mechanism and poses the same unreasonable question. Can it run AI?

The answer is almost always yes, if you have enough of them, enough time, and absolutely no regard for practicality. The math is real. The premise is absurd. Welcome to the book.

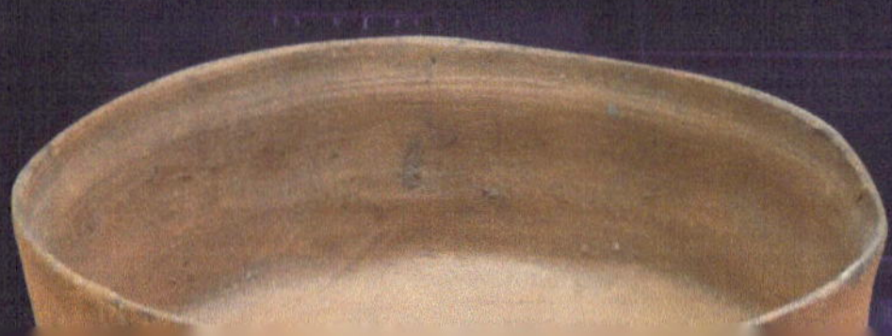

Run AI on Baghdad Batteries

betcha didn't know batteries came this old

250 BC
2,000+ year old battery

In 1936, German archaeologist Wilhelm König found clay jars in Iraq containing iron rods suspended in copper cylinders that are thought to be about 2,000 years old. The Baghdad Battery is magnificently simple. Just a clay jar, a copper cylinder, an iron rod, and an acidic electrolyte (vinegar, grape juice, or lemon juice in reconstructions). Fill one with vinegar and you get about 1.1 volts, enough to tingle your tongue if you're the kind of person who licks ancient artifacts. The iron oxidizes, releasing electrons that flow through an external circuit to the copper, which reduces. It's a galvanic cell, the same principle as your car battery, just 2,000 years older.

TECHNICALLY GALVANIC. COMPUTATIONALLY PALEOLITHIC.

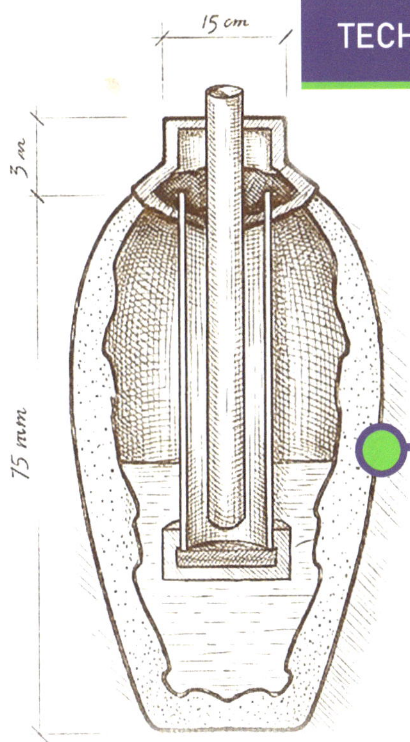

Can it Run AI?

With what: `vinegar battery`

Output: `galvanic direct current`

Unit: `per battery`

Watts: `0.0011W`

Yes, if you wire up 27,273 of them in parallel.

You cannot run inference on 0.0011W. The minimum power for even the tiniest model is around 1W. So wire them up in parallel (30W), and it is enough for one anciently electrified AI 'hello' in about 6 seconds.

So 1 battery = 11x Potatoes

and 1 cyclist = 90,909 Batteries

The Great Debate: Was it really a battery?

Archaeologists dispute if it was really a battery. The jars were found near Baghdad in 1936, and Wilhelm König later suggested they functioned as galvanic cells. Some also date them to Parthian or Sasanian period, so exactly how old they are remains debated.

~100 BC
2,000+ year old compute

Run AI on Antikythera Mechanisms

can we run AI on the world's oldest computer?

In 1901, sponge divers off Antikythera recovered a corroded lump from a Roman shipwreck. Inside were thirty bronze gears cut with triangular teeth, meshed in ratios that tracked the Metonic cycle, predicted eclipses, and modeled the Moon's elliptical orbit. It is the oldest known analog computer, built sometime around 100 BCE (give or take 30 years), and it worked. Someone in the Hellenistic world understood epicycles well enough to encode them in bronze, then lost the knowledge for two millennia. It is unquestionably a computer. It takes an input (crank position), performs a deterministic calculation (gear ratios encoding astronomical models). So, can it run AI?

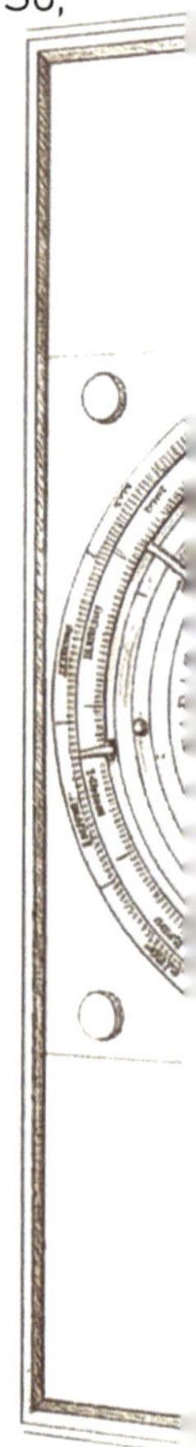

Can it Run AI?

With what: hand-cranking
Output: gear-rotation compute
Unit: operations per second (~30 gears×2 cranks/sec)
Compute: ~60 ops per second

Yes, in about 7.4 years.

You'd get a token in about 7.4 years, so your AI could say 'Hello.' But hey, no no subscription fee & it works underwater.

So 1 crank is producing about 15W, that's the power of about...

150,000x **Potatoes**

and 1 cyclist's power takes about...

6.7 **Cranks**

How it Works: The Antikythera does addition via gear ratios. It can't multiply directly.

The mechanism has approximately 30 meshed gears performing multiple linked analog calculations at once, but they're all working together on one compound astronomical operation, updating solar position, lunar phase, eclipse prediction, and calendar dials simultaneously. Brilliant at astronomy. Not great at matrix multiplication.

To simulate one multiply-accumulate for AI inference, you'd need to repurpose those gear meshes as computational steps. The mechanism has about 30 gears, each performing one operation per revolution. Crank at a comfortable 2 turns per second and you get roughly 60 operations per second. Impressive for 87 BCE. Less impressive when you learn what AI actually needs. A single forward pass through a small 7B parameter AI model requires approximately 14 billion multiply-accumulate operations. At 60 per second, that's about 7.4 years per token. Using the Antikythera for AI is like hiring an astronomer to do your taxes.

THE AI TOOK 7 YEARS TO SAY 'HELLO' & WAS ERGONOMICALLY DEVASTATING.

Each of the ~30 gears performs one operation per revolution. At 2 cranks per second, that's ~60 operations per second. So you'd need: 14,000,000,000 ÷ 60 = 233,333,333 seconds per token. That's 7.4 years of continuous cranking for one token to get to that first AI saying 'hello' to you.

You want a full sentence? You'd better start cranking now and circle back next decade with your kids.

12-1300 BC

3000+ year old clocks

Run AI on Water Clocks

reliable compute, drop-by-drop

The ancient Egyptians and Persians measured time with falling water. The Chinese measured it with burning incense. In both, the elegant part is that the measuring device consumes itself. The water that marked the hour is gone. The incense that counted the minutes is ash. Each computation destroys the computer that performed it. A water clock converts gravitational potential energy into timekeeping through calibrated flow. Water descends from upper to lower reservoir at a rate determined by orifice geometry and fluid dynamics.

Can it Run AI?

With what: `water clock`

Output: `falling water`

Unit: `clepsydrae (1.5L reservoir, 40cm drop, 1-hour cycle)`

Watts: `0.00163W per clepsydrae`

Yup, we need 18,405 parallel clepsydrae to get one AI to say 'hello'.

That gets us to 30W, and a 6 second time to 1st token.

So 1 clepsydra = 16x **Potatoes**

and 1 cyclist = 61,350 **Clepsydrae**

One clepsydra at 0.00163W cannot run AI inference. The power trickles, literally. You cannot accumulate falling water into compute any more than you can save yesterday's time.

The power available is modest, about 1.5 liters falling 40 centimeters per hour yields 0.00163 watts, assuming perfect conversion of potential energy to electrical work via micro-hydroelectric capture. In incense clocks, combustion releases 7.4 watts thermal, though Seebeck conversion at 5% efficiency yields only 0.37 watts electrical. Both methods share the fundamental constraint that the energy source is the timekeeping mechanism.

COMPUTE THAT DESTROYS ITSELF, IT CANNOT BE REBOOTED.

Discovered in 1904, the Clepsydra of Karnak water clock from about 1391-1353 BC, Pharaoh Amenhotep III.

The elephant water clock by medieval Islamic engineer Ismail al-Jazari in 1206 AD.

Run AI on Clock Towers

when the bell rings, an angel sings (in wattage)

1300 to present
700+ years of counting

The pendulum escapement is one of humanity's first truly discrete timekeeping mechanisms. At their peak there were as many as 10,000 clock towers with bells through geared clocks in Europe, of several varieties. A pendulum swings. A toothed wheel catches, releases, advances one position. Each tick is a discrete state change, analog motion converted into perfectly counted steps. A tower clock generates about 0.34 watts, relentlessly. The Westminster Clock has ticked over 2.5 billion times (that's about 21,600 swings/day) since 1859. No GPU can claim this uptime. This mechanism has kept time for 4 centuries without software updates or cloud dependencies.

89 CLOCK TOWERS FOR ONE AI CHAT. THE INQUISITION WOULD HAVE HAD QUESTIONS.

You'd need 89 towers to run AI inference. Medieval Europe was estimated to have 10,000. Nobody wired them together.

Can it Run AI?

With what: gravity & weights
Output: mechanical escapement
Unit: watts per tower
Watts: ~0.34W per tower (300kg, 10m, 24hr)

Sure, with 89 towers & luck.

Each tower delivers 0.34W continuous. 89 of them reach 30W inference. The Westminster Clock has maintained over 99% uptime for 167 years. (Better uptime than most SaaS products)

So 1 tower = 3,400x Potatoes

and 1 cyclist = 294 Towers

~1800
battery that started it all

Run AI on Voltaic Piles

could Napoleon have run AI?

In 1800, Alessandro Volta stacked zinc and copper discs with brine-soaked cardboard between them and created the first true battery. It powered the birth of electrochemistry and gave us the "volt" as a unit. Napoleon was so impressed he made Volta a count (Kipnis, 1982). What the pile did not do was generate meaningful power. A well-constructed Voltaic pile produces about 10 mW, enough to make a frog's leg twitch, barely enough to light a modern LED.

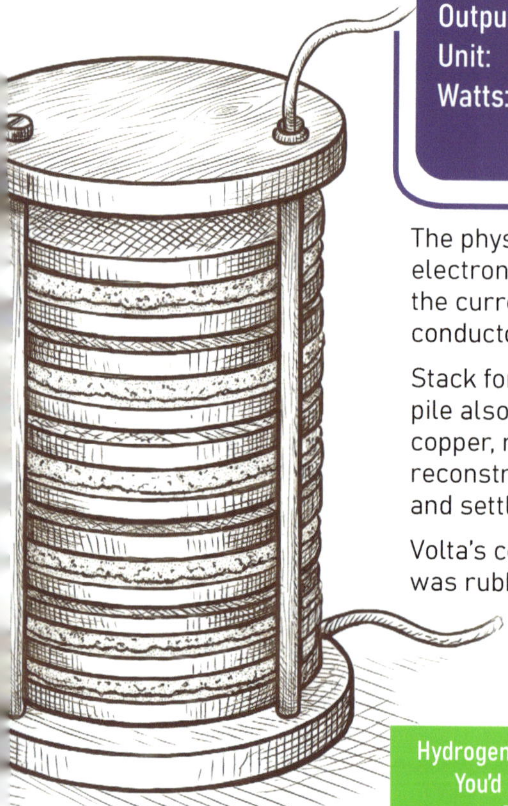

Can it Run AI?

With what: zinc/copper discs
Output: galvanic DC current
Unit: watts per pile
Watts: ~0.01W sustained (40 cell pairs)

Yea, connect ~3,000 piles.

Each pile sustains 10 mW after depolarization, wire 3,000 and you get 30W to an AI "hello."

1 Pile = 100x Potatoes
1 cyclist = 10,000 Piles

The physics is straightforward: zinc oxidizes, copper reduces, electrons flow. Each disc pair generates roughly 0.76 volts, but the current is tiny because brine-soaked cardboard is a terrible conductor.

Stack forty pairs and you get 30 volts at 0.3 milliamps. The pile also depolarizes as hydrogen bubbles accumulate on the copper, meaning it gets worse the longer you use it. Modern reconstructions at the Smithsonian peak at 50 milliwatts fresh and settle into a depressing 10 milliwatts sustained.

Volta's contemporaries were amazed because the alternative was rubbing amber with cat fur. Standards have evolved.

~10x more electricity than rubbing a cat (see also Cat Static)

Hydrogen buildup means cardboard breakdown. You'd need a cardboard rotation team. First data center job in history.

Run AI on Aeolipiles

Hero of Alexandria built a jet engine & nobody cared

100 CE
steam engines for parties

In the first century, Hero of Alexandria built a hollow bronze sphere that spun when heated. Steam from a boiling cauldron entered through hollow support arms and exited through two small opposing nozzles, spinning the sphere by jet propulsion. It was Newton's third law made visible seventeen centuries before Newton wrote it down (Stong, 1966). Hero used it to impress dinner guests. The Romans said "neat" and went back to using human power.

HUMANITY'S 1st STEAM ENGINE IN 100 CE WAS USED TO SPIN A BALL.

Can it Run AI?

With what: early steam turbine
Output: rotational power
Unit: watts per aeolipile
Watts: ~0.5W (well-built replica)

Sue thing, with 60 aeolipiles.

Half a watt from a boiling reservoir. Sixty running simultaneously reaches 30W. The fuel is wood or charcoal. So sixty open fires in a room becomes the challenge to get to 1 token.

So 1 aeolipile = 5,000x **Potatoes**

and 1 cyclist = 200 **Aeolipiles**

~250 BC to present
2,300+ year old tech still used

Run AI on Archimedes Screws

energy when you screw up, literally

The Archimedes screw has been lifting water since about 250 BC, a helical surface wrapped around a shaft, tilted at an angle, turned by hand or beast. Water climbs upward through the threads, defying gravity through geometry alone. Archimedes is traditionally credited with the screw and may have designed it to drain bilge water from ships, though the historical record is vague about whether he was solving a problem or showing off (or if he adapted from older designs). We propose running it backward, letting falling water or hay or whatever really, spin the screw to generate electricity for inference.

The Archimedes screw is still used today in tractors and harvesters.

Most chapters in this book end in millions of potatoes and quiet despair. The Archimedes screw joins a short list that genuinely crosses the 30W threshold.

One micro-hydro screw at a modest stream produces 8,000W. That's enough to power 266 quantized 7B models running real-time conversation simultaneously, all day, every day, as long as the water flows. One model needs 30W continuous. The screw barely notices. Fish-friendly too — the open helix lets aquatic life pass through unharmed.

Can it Run AI?

With what: `reversed helical screw`

Output: `hydroelectric power`

Unit: `watts per screw`

Watts: `~8,000W (3m dia, 2m head, 2.5 m³/s)`

Yes, this actually works.

One micro-hydro screw at a modest stream produces 8,000W, enough you could feasibly run modern AI all day. And have happy fish too.

So 1 crank is producing about 15W, that's the power of about...

80,000,000x
Potatoes

and 1 cyclist's power takes about...

1/80th (0.00125)
Screws

The Archimedes screws are also used in dams.

Similar screw technology is found inside large commercial chocolate fountains to lift chocolate to the top.

How it Works: We put energy in to move water up, put water in to get energy out

Reversed, an Archimedes screw becomes a hydroelectric generator. Water flows down through the helical channels, spinning the screw, which drives a generator. Modern micro-hydro installations at small dams and irrigation sites achieve 80–85% efficiency, better than conventional turbines at low head heights (Fishtek, 2012). A typical small-scale screw (3m diameter, 5m long, 2m water drop) produces roughly 8,000 watts.

Eight thousand watts. That's 266 simultaneous AI conversations from one creek and a 2,300-year-old invention. No exotic materials. No combustion. No subscription. Just water falling through a very clever spiral that predates the printing press by seventeen centuries and is still manufactured, largely unchanged, worldwide.

This is the chapter where the book's premise breaks entirely. Archimedes solved your infrastructure problem before Rome fell.

AI Can Improve Archimedes Screws by 60-70%

AI has been used to turn hydraulic machines into a smart, efficient, and sustainable technology for energy production and water management. And Micro-hydro Archimedes Screw Turbines (ASTs) really are incredibly fish-friendly.

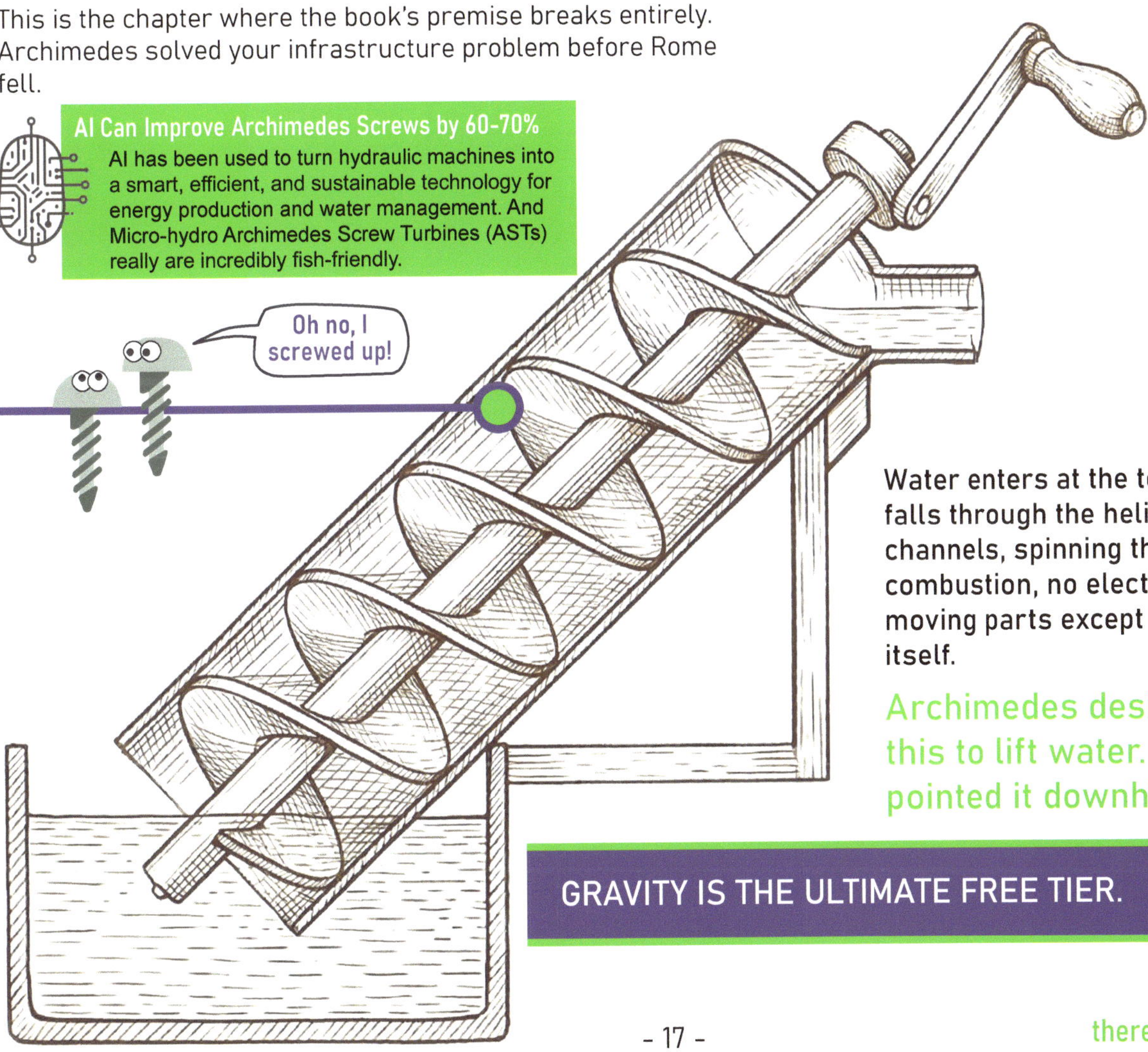

Water enters at the top and falls through the helical channels, spinning the shaft. No combustion, no electronics, no moving parts except the screw itself.

Archimedes designed this to lift water. We just pointed it downhill.

GRAVITY IS THE ULTIMATE FREE TIER.

Section 2

MECHANICAL CALC

Machines that actually do math perfectly fine, just very, very slowly.

Every device in this section performs genuine mathematical operations. Not metaphorical math. Not "it generates a voltage that could theoretically represent a number." Actual multiplication, actual counting, actual computation, etched into gears, carved into stone, slid along calibrated wood, pressed into wet clay.

The pinwheel calculator multiplies through interlocking teeth. The slide rule adds logarithms by physically moving one scale against another. The Jacquard loom reads binary instructions from punched cards while making scarves. The brass astrolabe solved trigonometry for centuries and would need all of them again to make AI say 'hello,' and a Hammond B-3 has been performing real-time analog computation since 1935.

Some of these chapters measure time. Some measure power. One measures how loud 6,000 desk toys get when you refuse to give up. The answers range from years to millennia, from milliwatts to chainsaw. Bring a book. Preferably this one.

THE JACQUARD LOOM is the direct ancestor of modern computing. Hollerith adapted the punched card for the 1890 US Census. IBM was originally a punched card company. The card format survived until the 1970s. The conceptual leap (storing instructions on a physical medium, reading them sequentially, executing them mechanically) is exactly what your GPU does. Just 56 trillion times per second instead of once.

Run AI on Mechanical Looms

you get power, and a rather nice silk scarf

1804 to present
1st programmable silk machine

In 1801, Joseph Marie Jacquard built a loom that read punched cards to weave patterns automatically. Each card was a line of binary: hole or no hole, thread up or thread down, one or zero. The loom was programmable. Weavers rioted. Its punched-card logic later influenced Charles Babbage's Analytical Engine and the computing ideas Ada Lovelace wrote about. The cards became the first software, stored in stacks, debugged by hand, versioned by sequence number. We propose using it for inference. The output layer produces a scarf.

THE FIRST VERSION CONFLICT WAS A TANGLED THREAD.

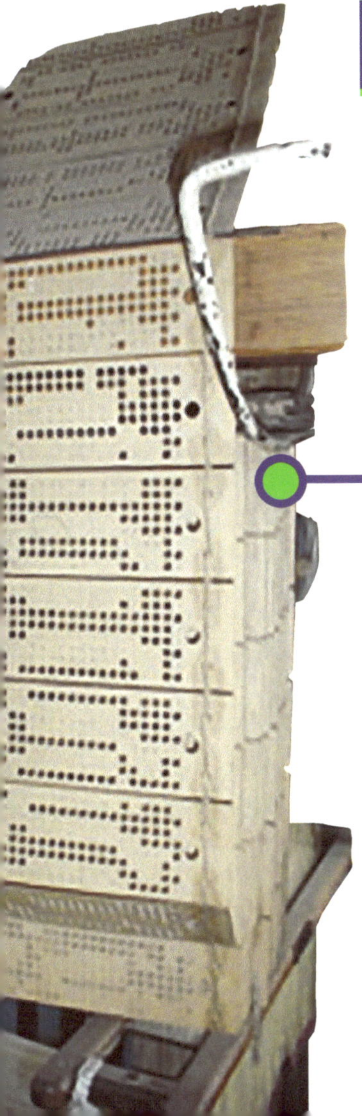

Punched cards advance on a chain, one per pass. Spring-loaded hooks probe each position, hole means the thread lifts, no hole means it stays down. A single card controls 400 threads simultaneously: 400 parallel binary operations in wood and wire. At one card per second, the loom performs about 25 effective multiply-accumulate operations per second. Saying 'Hello' takes 89 years and 2.8 billion cards chained end to end, a ribbon of instructions stretching 452,000 miles, nearly twice the distance to the Moon.

Can it Run AI?

With what: `Jacquard punched card`
Output: `parallel binary (400 threads)`
Unit: `effective ops per second`
Compute: `~25 ops/sec (400 binary ops/card, 1 card/sec)`

In 89 years, with 452,000 miles of punch cards.

Each card executes 400 simultaneous binary decisions, compute in wood and metal. At this rate, one token takes 18 years and one 'Hello' takes 89 years, and the punch card reaches the moon & back.

So 1 loom = 600,000x Potatoes

and 1 cyclist = 1.7 Sleepless Weavers

1673-1930s
1st serious mechanical compute

Run AI on Pinwheel Calculators

compute without a cloud subscription required

This is a Rapid Calculator, built in the United States around 1890. It is not an adding machine, it is a calculating machine that multiplies and divides with a Leibniz wheel (or stepped drum), a brass cylinder with nine teeth of increasing lengths that was invented by Gottfried Wilhelm Leibniz around 1671–1673. It's part of a long history from Babbage to the modern GPU.

You set a number on the rotating drum. Crank the handle. It adds. Shift one position, crank again, you've just multiplied by ten. That's multiplication through addition with positional shifting. The same operation that modern neural networks in AI perform on GPUs billions of times per inference.

For two centuries, these machines did the math that built railroads, balanced ledgers, and calculated artillery tables.

THIS MACHINE DOES THE RIGHT MATH FOR AI, JUST NOT ENOUGH OF IT.

How it Works: The Pinwheel does multiply-accumulate natively.

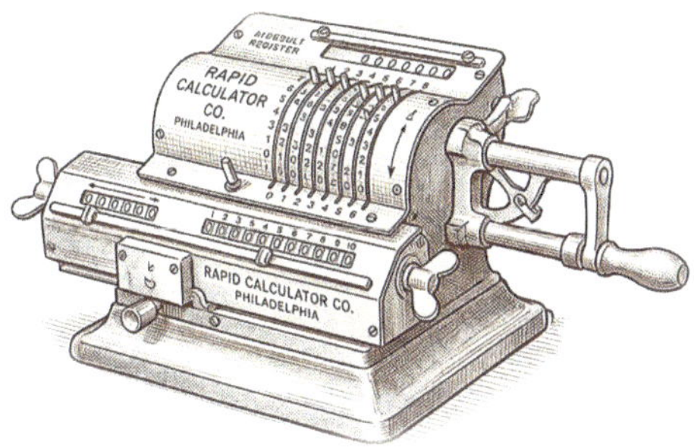

That's literally what the Leibniz wheel was invented for. One crank = one multiply-accumulate. It's doing the exact operation transformers need. An experienced operator could perform roughly 2 complete multiply operations per second, factoring in digit setting, carriage shifts, and the occasional rest for your increasingly cramped hand.

It does the right math. But if you'd started cranking at the fall of Rome, you'd be finishing up about now. Every operation would be correct. No floating point errors. No hallucinations. The slowest, most accurate inference engine ever built.

Not Just Power- It computes!

While its calculations may be slow, they are the same type used in modern compute. Just needing more elbow grease. A lot more.

Rapid Calculator. Otto P. Schuman (1891-1969) arrived in the United States from Germany in 1909 and soon after founded the Rapid Calculator Company in 1919, applying for calculating patents that same year (1,508,257 - granted September 9, 1924) and 1924 (1,566,978 - granted December 22, 1925).

Can it Run AI?

With what: Leibniz wheel

Output: mechanical multiplication

Unit: operations per second

Compute: ~2 ops per second (experienced operator)

Yes, it gets to say 'hello' in 1,110 years.

A small 7B parameter AI requires 14 billion multiply-accumulate operations per token. The Leibniz wheel does this, but slow. At 2 per second, that's 7 billion seconds. 222 years for your AI to say 'Hello' to you.

So 1 crank is producing about 15W, that's the power of about... 150,000x Potatoes

and 1 cyclist's power takes about... 6.7 Cranks

You Started Cranking at the Fall of Rome
our AI is now ready to say 'Hello' to you.

1829 to present
for the DIY crowd in the room

Run AI on No. 2 Pencil & Paper

if you feel like doing it by hand

The modern No. 2 pencil was developed in the early 19th century, though some sources say true prototypes of the modern graphite-clay pencil were introduced as early as 1795. Inventor Joseph Dixon created a precursor in 1829 using graphite, while Henry David Thoreau perfected the mixing of graphite and clay for varied hardness in the 1820s. The iconic yellow Dixon Ticonderoga No. 2 was formally introduced in 1913.

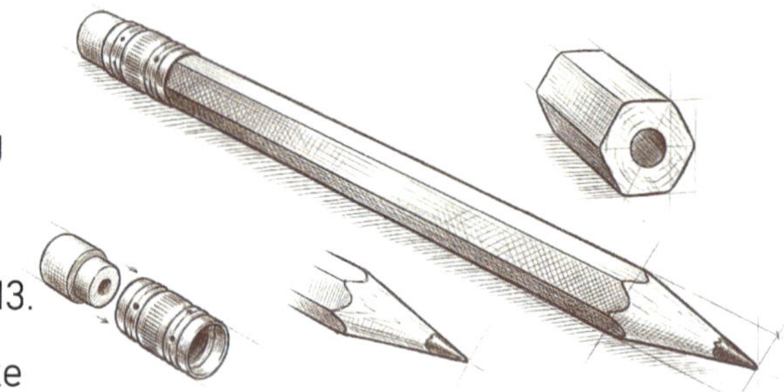

For the DIYers in the room, how long do you think it would take to manually calculate your AI's inference?

YOUR BRAIN IS THE GPU. THE PEN IS THE BUS. THE PAPER IS RAM.

Can I Run AI on Paper?

With what: pen and paper

Output: Arabic multiplication

Unit: operations per hour

Compute: ~600 operations/hour (skilled human)

Yes, in ~2,663 Years.

You'd consume about 175 tons of paper (140,000 reams, stacked 4.3 miles high) to calculate your first AI 'Hello.'

That means each human has power of... 150,000x Potatoes

and 1 cyclist has the power of ~6.7 humans with pencils

What About in Stone?

With what: clay tablet + stylus

Output: Cuneiform computation

Unit: operations per hour

Compute: ~60 operations/hour (skilled scribe)

Yes, in ~26,600 Years.

A 7B model needs 14 billion multiply-accumulate operations for 'Hello'. That's 233 million hours & ~187 million stones.

That means each human has power of... 25x Potatoes

and 1 cyclist has the power of ~40,000 scribes with stone

* Based on metabolic power 15W, 150,000 potatoes, ~6.7 pencil-computers per cyclist

Run AI on Stone Tablets

the best guaranteed uptime, but long wait

3000 BC
a.k.a 5000+ year old stone

PERMANENT. ILLEGIBLE TO FUTURE ARCHAEOLOGISTS. EXTREMELY DURABLE.

Clay tablets of ancient Mesopotamia have survived five millennia. They outlasted empires that created them, the languages that read them, and storage medium invented since.

Your cloud backup has a Service Level Agreement (SLA) for uptime. Cuneiform has an SLA measured in civilizational collapse. But can that wet clay compute an AI inference one operation at a time?

And Good News, It's Archival Quality Data Storage

Babylonian scribes were mathematicians. They had multiplication tables pressed into reference tablets, solved quadratic equations, and tracked compound interest. They could compute. The problem isn't the brain, it's the medium. Writing one number in cuneiform requires 3 to 8 separate wedge impressions, each pressed carefully into wet clay. Recording a single multiply-accumulate result takes about a minute. At 60 operations per hour, a trained scribe was the fastest computer in 3000 BCE, computing using only careful, deliberate, and measured in the pressure of human hands against clay.

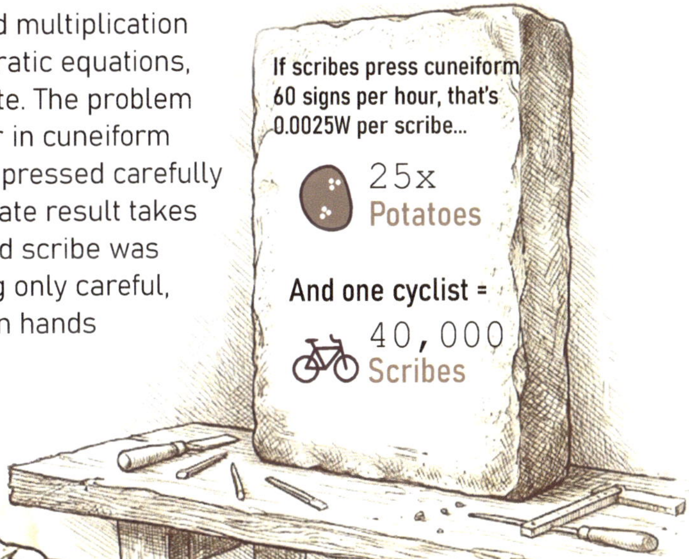

AI IS REVOLUTIONIZING ARCHAEOLOGY

Modern AI can identify translate, and reconstruct ancient cuneiform, Greek, and other stone tablets. Models like ProtoSnap and DeepMind's Ithaca & Pythia AI can even decipher broken fragments and determine the rules of ancient games

Computational-Equivalent Wattage of Stone

If you convert a scribe's computational output to its electrical equivalent, they generate about 0.0025 watts of useful compute. That's less than a hearing aid. The other 79.9975 watts are posture, frustration, and clay management.

1632-1972
350 years of engineering

Run AI on Slide Rules

every engineer who got us to the moon used one

The slide rule dominated computation for three centuries. Groundwork from logarithms was first published in 1614 and Gunter's scale before. Then in 1632, English mathematician William Oughtred designed the first adjustable logarithmic rule. Engineers designed bridges with them. Physicists split atoms. NASA went to the moon. The device multiplies by physically adding logarithmic distances, slide one scale against another, read the result where they align. No batteries, no software, no cloud outage. A skilled operator completes one multiplication in about four seconds. At that rate, saying "Hello" takes 1,775 years. A tournament of 100 operators in parallel brings it down to 18 years, still longer than a career, but technically achievable if anyone can maintain the bracket schedule that long.

Can it Run AI?

With what: calibrated log scales

Output: analog multiplication

Unit: operations per second

Compute: ~0.25 ops/sec (skilled operator)

It'll only take about 1,775 years if the NASA engineers work in shifts.

One operator, one slide rule, four seconds per multiply. Math checks out, but with 100 operators it takes 18 years.

So 1 slide rule = 200,000x Potatoes

and 1 cyclist = 5 Operators of Slide Rules

YOUR GPU IS ~56 TRILLION TIMES FASTER.
* Compared to modern high-end GPU at 14 TFLOPS

OMG, the Slide Rule is Making Us Dumber!

Resistance to AI in 2020s closely mirrors mid-1990s "calculator panic" against handheld calculators and slide rules, where many feared human cognitive decline from machine use. This includes famous 1986-88 protests of the handheld calculator by math teachers in Washington D.C.

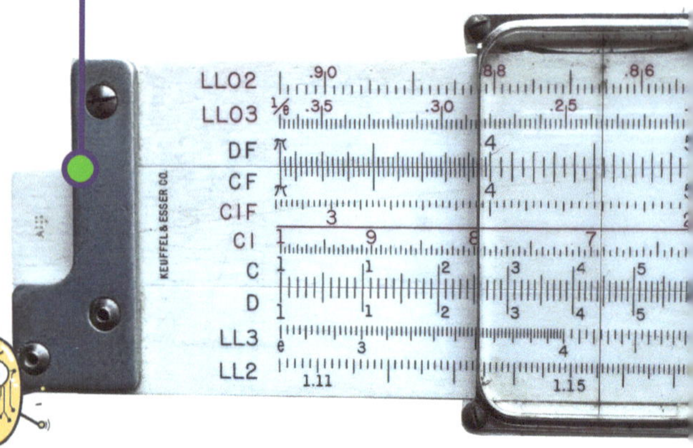

Otis King's Patent Calculator (the cylinder)

Patented 1921 by Otis Carter Formby King, made by Carbic Limited, London. The helical scale wraps around the cylinder giving roughly 66 inches of effective scale length in a pocket-sized tube. More precise than a standard 10-inch rule by about 6x. Marketed as a pocket calculator. It was not lying. "British Made Throughout" was not marketing.

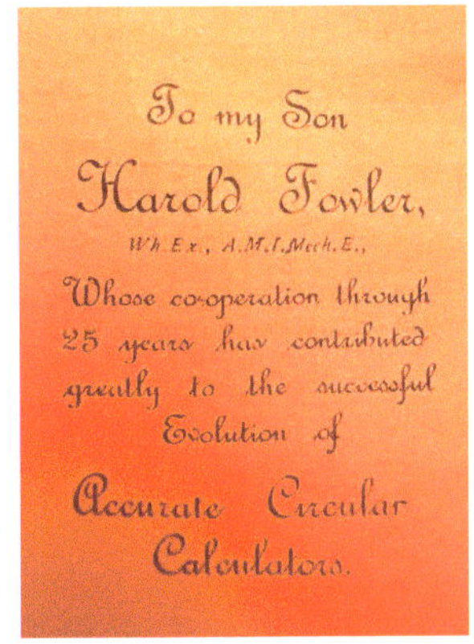

Fowler's Universal Calculator (the circle)

Circular slide rules pack longer scales into smaller diameters through concentric rings. Fowler's Manchester operation made dozens of variants, the Universal was their flagship. The disc format was more compact and elegant than straight rulers. Engineers who owned both usually had opinions about which was superior. They were wrong about whichever one they preferred less. And, fun fact, he dedicated the design to his son.

TWO BRITISH INSTRUMENTS. TWO PHILOSOPHIES.

These devices function like a slide rule, but using a cylinder or a disc. Both solve the same problem of getting more scale in less space through completely different geometry. Both are now archival quality data storage for the specific kind of data nobody needs anymore.

1887-1960s
it still beats the abacus

Run AI on Adding Machines

for when you don't have enough fingers

The adding machine does exactly one thing. It adds. Pull a lever, dial with a stylus. One number joins another, a running total advances, the carriage returns. Multiplication is repeated addition with a shift. The machine that could only add is the ancestor of the machines that can do everything. A skilled operator does roughly 0.33 multiply-accumulate operations per second through shift-and-add. At that rate, saying 'Hello' takes about 1,345 years. A typing pool of 100 lowers it to about 13 years.

Can it Run AI?

With what: mechanical key-driven

Output: analog sum via gears

Unit: operations per second

Compute: ~0.33 MACs/sec (skilled operator, shift-and-add)

Sure, in about 1,345 years.

One operator, one machine, nine additions per multiply. The math works. The timeline doesn't. But the 10-key layout survived into every GPU data center on Earth. The keys won.

So 1 adding machine = 200,000x Potatoes

and 1 cyclist = 5 Operators

Burroughs Class 3

William Seward Burroughs patented the first practical adding machine in 1888, then founded the company that would become Unisys. This one has a dedicated ERROR key. Maybe it's trying to warn you it still adds faster than you do.

ADDING MACHINES ARE GOOD FRIENDS. YOU CAN ALWAYS COUNT ON IT.

Run AI on Astrolabes

astronomically precise, computationally glacial

150–225 BC
1st portable computer

The astrolabe is humanity's first smart computer. Invented in Hellenistic Greece, perfected in medieval Damascus, it solves spherical trigonometry through the physical rotation of engraved brass plates. The finest medieval examples held tolerances under 0.1 degrees. Align the rete with a star, rotate the alidade to the horizon, read your latitude from the scale. No calculation required, the mechanism is the calculation. For eight centuries it was one of the most sophisticated instruments a person could own. The astrolabe predates written language by less time than it would need to finish one conversation. We propose using one to run your AI.

THAT'S LONGER THAN THE ASTROLABE HAS EXISTED. THE BRASS IS PATIENT.

PARTS OF THE BRASS ASTROLABE

An astrolabe has three main parts: the mater (base plate with degree markings), the rete (rotating star map), and the alidade (sighting rule). Each rotation of the rete against the mater performs an analog trigonometric calculation encoded directly into the brass gear ratios. The finest medieval examples, like those from 13th-century Isfahan, held precise tolerances. They ran on human curiosity and the occasional drop of oil.

Some Historians Call it a "Medieval Smartphone"

Used from the 6th century through the Middle Ages, the astrolabe held over 400 different functions.

Can it Run AI?

With what: brass & mater plates
Output: spherical trigonometry
Unit: operations per second
Compute: ~0.15 ops/sec (continuous rotation)

Sure, in about 2,958 years

Each rotation solves one trigonometric position. At 0.15 operations per second, one token takes 2,958 years, and you finally get that first sweet, sweet AI 'Hello.'

So 1 brass = 50,000x Potatoes

And 1 cyclist = 20 Astrolabe users

1935 to present
a century of analog compute

Run AI on Hammond Organs

inference could hum in B-flat, maybe on key?

The Hammond B-3 does not generate tone acoustically. It generates tone through ninety-one ferrous tonewheels spinning past electromagnetic pickups at precisely 1200 RPM, generating sine waves that combine into what the patent called "electrical musical tones." Each wheel has between 2 and 192 teeth machined into its rim. Pull the drawbars and you're not opening valves, you're setting Fourier coefficients, mixing electromagnetic waveforms at the bus bar. The organ has been performing real-time additive synthesis since 1935, genuine analog computation, ninety-one parallel channels, phase-locked to the power grid. The organ produces roughly 2 watts of harvestable electromagnetic signal after the preamp stage. It consumes 350 watts to get there. Efficiency: 0.57%. The rest becomes heat, vibration, and that specific 60 Hz hum every jazz organist knows is in tune.

Can it Run AI?

With what: electromagnetic output

Output: additive sine synthesis

Unit: full organs (all drawbars pulled)

Watts: ~2W per organ (signal only)

15 Organs, drawbars out.

Playing simultaneously they produce 30W of harvestable electromagnetic signal. The sound is magnificent. The power efficiency would be criminal, consuming 5,250W of mains power to get to 30W for tokens.

So 1 organ has the power of about...

20,000x Potatoes

and 1 cyclist's has the power of about...

50 Hammond organs

HAMMOND SOUND GENERATED WITH AI

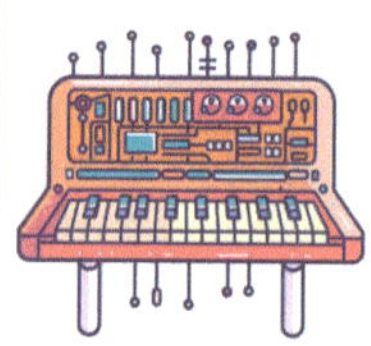

AI is increasingly used to simulate the iconic electromechanical sound of the Hammond B3 organ, the 1935 instrument best know for its role in jazz and gospel music for "soulful and grittty" sounds from spinning metal wheels and electromagnetic pickups. IK Multimedia, Sonauto and Suno others use AI-driven modeling to generate Hammond B-3 style sound.

Run AI on Quartz Oscillators

your watch knows what time it isn't

A quartz crystal oscillates at 32,768 Hz when voltage runs through it, so stable we built civilization's timekeeping on it. The piezoelectric effect works both ways. Apply voltage, crystal deforms; deform crystal, get voltage. A standard watch crystal consumes about 1 microwatt to maintain oscillation. That's 0.01 potatoes. One hundredth of a potato. Thirty million crystals vibrating simultaneously would reach 30 watts, roughly a bathtub full of quartz. But then there's cesium. A cesium-133 atomic clock measures time by counting 9,192,631,770 hyperfine transitions per second, which is how we define the second. The catch is that a laboratory cesium fountain consumes 200 watts of lasers, microwave interrogation, and magnetic shielding to observe those transitions. It doesn't generate power. It consumes it to tell you precisely how long your AI isn't happening.

1927 to present
still used in precision clocks

ATOMIC CLOCK:

158 MILLION YEARS AND STILL RIGHT

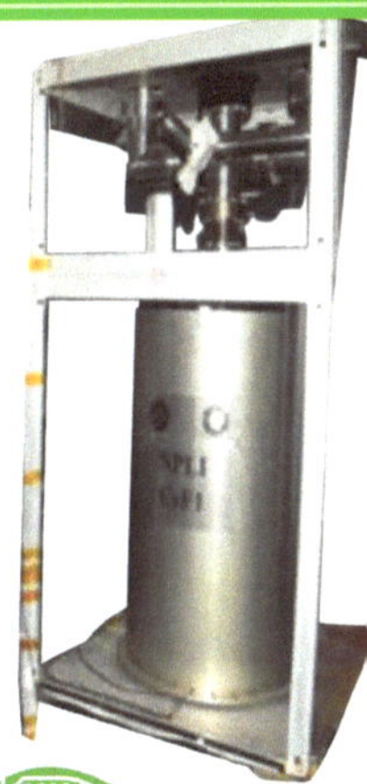

A cesium atomic clock is a master timekeeping device so precise its only loses 1 second every 158 million years, meaning they would be accurate since the Jurassic period.

INSANE ACCURACY

A quality quartz clock can keep time within 1 second over 30 years. Most quartz clocks and watches use a crystal cut to vibrate at exactly 32,768 Hz (or 2^{15} times per second) because this specific frequency is easy to divide down to exactly 1 pulse per second.

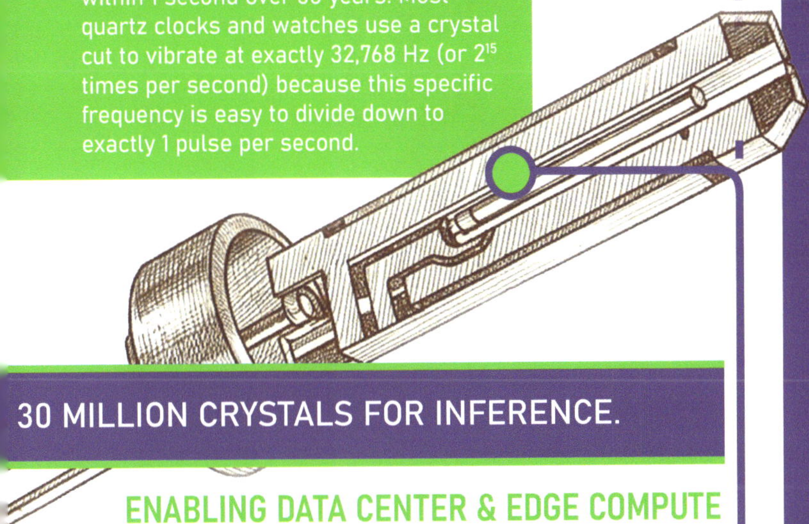

30 MILLION CRYSTALS FOR INFERENCE.

ENABLING DATA CENTER & EDGE COMPUTE

Quartz oscillators are used as high-precision timing components in AI systems, with stable, low-noise clock signals used in processors critical to AI data centers, 5G base stations, and edge devices. They enable AI training and inference.

Can it Run AI?

With what: piezoelectric crystal oscillation

Output: vibration energy

Unit: watch crystals (32.768 kHz)

Watts: ~0.000001W (1 µW) per crystal

Just need 30,000,000 crystals.

A bathtub of almost 30 million crystals reaches 30 watts, enough to say "Hello" in about 6 seconds. An atomic clock burns 200 watts just to confirm it took 6 seconds.

So 1 crystal = 0.01x Potatoes

and 1 cyclist = 100,000,000 Crystals

1967 to present
and still an amazing desk toy

Run AI on Newton's Cradles

the desk-friendly conference call power source

The Newton's cradle has graced boardroom desks since 1967, clicking through conference calls as a reminder that physics exists and meetings do not. Lift the first ball, release. Potential energy becomes kinetic. Momentum transfers through three stationary balls to the fifth, which swings outward. The process reverses. Each collision maintains 96–97% efficiency, which sounds impressive until you realize the remaining 3% is why it stops after thirty seconds unless someone gives it another push.

All balls in a Newton's cradle need to weigh the same. Because momentum and kinetic energy depend on mass, unequal balls break the conservation laws that make the cradle work, causing the balls to fly off at different speeds and heights

Most chapters in this book require thousands or millions of something. Newton's cradle requires 6,000. That and a soundproofed room.

Each cradle produces 0.005 watts with manual resetting. 6,000 clicking in unison reach 30W, enough for one AI conversation. The sound reaches 103 decibels, roughly a chainsaw. The momentum is conserved. Your hearing is not. Conference call optional but not recommended.

Can it Run AI?

With what: steel ball collisions

Output: kinetic energy transfer

Unit: cradles (manually reset)

Watts: ~0.005W per cradle

Yup. 6,000 cradles is all.

Each cradle generates 0.005 watts, about 50 potatoes' worth. 6,000 clicking in unison reach 30 watts, enough to say "Hello" in 6 seconds. Each reset requires a human finger. The momentum is conserved. Your afternoon is not.

So 1 cradle =

50x Potatoes

and 1 cyclist's power takes about...

20,000 Cradles

Typical Newton's cradle balls for desktop models are made of steel, with individual ball weights often ranging from roughly 20 to 100+ grams, depending on the size of the cradle. High-quality, larger, or heavy-duty units may feature individual balls weighing over 100 grams maximize momentum.

20 to 100+ GRAMS

FAQ: Every Wonder how Loud 6,000 Running at Once Would Be?

A standard desktop cradle generates about 0.005 watts with continuous manual resetting, the power output of someone who's given up but hasn't admitted it yet. You'd need 6,000 synchronized cradles for inference. At 65 decibels per click, 6,000 in unison produce 103 decibels, the volume of a chainsaw. The inference would work. The boardroom would not.

How it Works: Conservation of Momentum, But a Long Way to Inference

Newton's cradle demonstrates conservation of momentum through elastic collisions. When the first ball strikes, it stops, transferring its momentum through the stationary middle balls as a compression wave in the steel, arriving at the last ball in about 0.0002 seconds. The last ball launches outward at nearly the same speed the first one arrived. Energy loss comes from air resistance, the audible click, and microscopic deformation at each contact point.

Why did the AI refuse to play fetch?

It was afraid of losing its tokens and couldn't handle the balls!

Louder than a Chainsaw

Each collision is 96% efficient. Each reset requires a human finger. 6,000 cradles produce 30 watts and 103 decibels, louder than a chainsaw.

MOMENTUM CONSERVED. MEETINGS DESTROYED. INFERENCE LOUD.

Section 3

RETRO TECH

The stuff in your parents' junk drawer and how it almost powered AI.

Everything in this section arrived within living memory, and most of it still works. The TI-82 graphed your parabolas in 1993 and somehow still costs about $100 because the American education system refuses to move on. The Nokia survived walls, toilets, and bullets. Punch-card computers filled rooms.

The gap between devices that take "thousands of years for AI to say hello" and "about 40 minutes" happened in roughly fifty years. Your parents and grandparents watched it happen. Some of them caused it. Some of them still have those old Nokia phones and Game Boys in junk drawers, and that's more compute than all prior centuries of human history combined. There is more forgotten compute in one American garage than existed on Earth in 1900.

Run AI on TI-82 Calculators

spending generations powering high schools

1993 to present
AAA battery's best friend

TI-82 has survived every wave of consumer electronics since 1993 (smartphones, tablets, cloud computing, the entire internet) and it's still in backpacks. The Zilog Z80 processor runs at 6 MHz with 28 kilobytes of RAM, powered by four AAA batteries at 50 milliamps. It draws 0.15 watts. It has no floating-point unit, no GPU, no network connection, and somehow the TI-whatever series still costs about $100 just like it did in 1996 because Texas Instruments has a monopoly on high school math and sees no reason to change. The Z80 achieves roughly 85,000 useful multiply-accumulate operations per second, with each multiplication requiring six or more instructions that a GPU handles in one cycle. Saying "Hello" takes 1.9 days. The batteries last two. You'll need two sets of AAAs, a steady hand on the ON button, and the kind of patience that only high school graphing homework could have prepared you for.

Can it Run AI?

With what: `Zilog Z80 @6 MHz, no FPU`

Output: `8-bit integer operations`

Unit: `operations per second`

Compute: `~85,000 ops/sec at 0.15W`

Sure. In 1.9 days.

The Z80 can technically do the math. Each multiply-accumulate operation just takes six instructions instead of one. The real problem is RAM: the model needs 4 gigabytes, the calculator has 28 kilobytes. We're being generous and ignoring this.

So 1 TI-82 = 1,500x Potatoes

and 1 cyclist = 667 Calculators

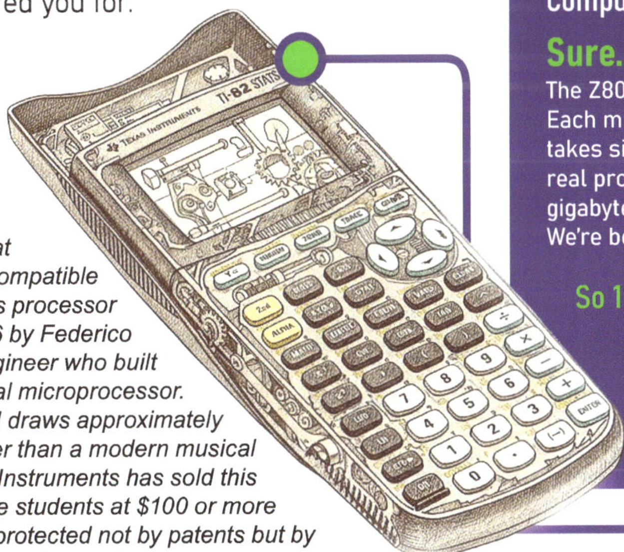

Zilog Z80

Toshiba T6C79 is a specialized ASIC (Application Specific Integrated Circuit) that incorporates a Z80-compatible core. This calculator's processor was designed in 1976 by Federico Faggin, the same engineer who built Intel's first commercial microprocessor. Its runs at 6 MHz and draws approximately 0.06 watts, less power than a modern musical greeting card. Texas Instruments has sold this architecture to captive students at $100 or more for over thirty years, protected not by patents but by standardized testing requirements.

A quantized 7B model needs 4 gigabytes of RAM. The TI-82 has 28 kilobytes. You'd need to swap data from external storage about 150,000 times, taking months.

THE AI MODEL DOESN'T FIT IN RAM BY A FACTOR OF ~150,000.

2001 to present
pretty sure it's immortal

Run AI on Nokia Brick Phones

this early 2000s staple is indestructible

The Nokia 3360 runs on an ARM7TDMI processor at 33 MHz, which was so efficient it went on to power the Game Boy Advance, the iPod, and half of the embedded systems on Earth. Unlike the TI-82's 8-bit Z80, the ARM7 has hardware 32-bit multiply. That means that one instruction does what takes the calculator six. Rated at roughly 33 million instructions per second, though each multiply-accumulate takes about five cycles and performs about 6.6 million operations per second.
So AI on the Nokia says "Hello" in about 40 minutes.

Founded in 1865, Nokia started as a paper mill in Finland

The name "Nokia" is derived from the town of Nokia, Finland and nearby Nokianvirta River.

As a bonus, it shipped with an amazing copy of Snake II, a flashlight, and the ability to survive a fall from any height. It drew 1 watt during active use and was, by every metric except camera and WiFi, a more reliable computer than whatever you're holding now.

2G

T-Mobile was among the last major U.S. carriers with lingering 2G, bits been gradual shutdown since 2025.

Secretly Bulletproof

The Nokia 3360 is a, compact candybar or "brick" style mobile phone released by Nokia in early 2001, specifically designed for the U.S. market using TDMA/AMPS technology. It is a variant of the popular 3310 series. Known for its durability, YouTubers have shown it survive a 9mm bullet shot, a hydraulic press, and more.

Can it Run AI?

With what: `ARM7TDMI @ 33 MHz`

Output: `32-bit hardware multiply`

Unit: `operations per second`

Compute: `~6.6M operations/sec`

Yes, 40 minutes, 1 charge.

What took the TI-82 days takes the Nokia only 40 minutes. The RAM shortfall is still absurd (200 kilobytes versus 4 gigabytes) but less absurd than the calculator's. We're still ignoring memory swapping. We could definitely make an AI talk on Nokia.

So 1 Nokia...

10,000x
Potatoes

and 1 cyclist's power takes about...

100
Nokia Bricks

40 minutes. On one charge.

The ARM7's hardware multiply changes everything. What took the TI-82 more than a day takes the Nokia 40 minutes. The RAM shortfall is still absurd (200 kilobytes versus 4 gigabytes) but less absurd than the calculator's by a factor of seven. We're still ignoring memory swapping. The inference would work. The phone would survive.

THE AI-NATIVE NOKIA

Nokia & NVIDIA are partnering to add AI-native capabilities to 5G-Advanced and 6G networks.

Why are they so strong?

The 3310 was designed with a tough, double-shell polycarbonate casing, a recessed screen, and a solid, non-glued construction that allowed it to absorb impacts, unlike modern glass phones.

SNAKE II RUNS FLAWLESSLY A DECADE AFTER 2G

Snake II for Nokia, famous for its inclusion on the Nokia 3310 in 2000, was developed in-house by Nokia. While Taneli Armanto programmed the original Snake for the Nokia 6110 in 1997, Snake II was part of a suite of games developed by Nokia just for their phones.

The Nokia 3360 runs on 2G GSM. AT&T shut down their 2G network in 2017. T-Mobile followed. The phone boots, holds charge, and plays Snake II flawlessly in 2026. It just can't reach a single cell tower on the continent. The hardware outlasted its own infrastructure. It has no WiFi. No internet. Only Snake and a handful of games. It was enough.

THE PHONE STILL WORKS. THE NETWORK DOESN'T.

The Nokia 3360 was bought from eBay in February 2026, listed as "not tested." It powered on immediately. The original battery held charge. Snake II loaded without hesitation. The original user manual was included, still sealed, with stock photography of people who are now in their forties. A 14-year-old field tester was given the device for evaluation. Within one hour he had discovered Snake II has levels, asked where the apps were (there are no apps), asked how to connect to wifi (there is no wifi), and located a loophole in his school's "no smartphones" policy, technically correct, as the Nokia 3360 is categorically, legally, and philosophically not a smartphone. Declared a "dumb phone" definitely, he concluded he could compliantly smuggle it to school. He has not returned it.

We powered on the Nokia 3360 untested from eBay. Two minutes of charge and we were playing Snake II like it's 2001. It's still fun.

1977-1992

128 bytes and a big dream

How long to AI "Hello" on game systems?

System	Year	Time to AI Hello
Atari 2600	1977	1.1 days
NES	1983	17.3 hours
SNES	1990	6.7 hours
Sega Genesis	1988	2.0 hours
PS1	1994	20.4 min
Sega Saturn	1994	12.0 min
N64	1996	5.0 min
Dreamcast	1998	2.3 min
PS2	2000	1.2 min
GameCube	2001	43 sec
Xbox	2001	19 sec

Run AI on Atari 2600s

woodgrain computer that defined a generation

The Atari 2600 has 128 bytes of RAM. Not kilobytes. Bytes. The MOS 6507 runs at 1.19 MHz with no multiply instruction. Every multiplication is a loop of shifts and additions taking eight or more clock cycles. The processor was designed to move sprites across a television, not to multiply matrices. It has no operating system. It has no storage.

The game cartridge is the program, mapped directly into the address space because there isn't enough RAM to copy it anywhere. To run a quantized 7B model you'd need 31 million times more memory than exists on the entire board.

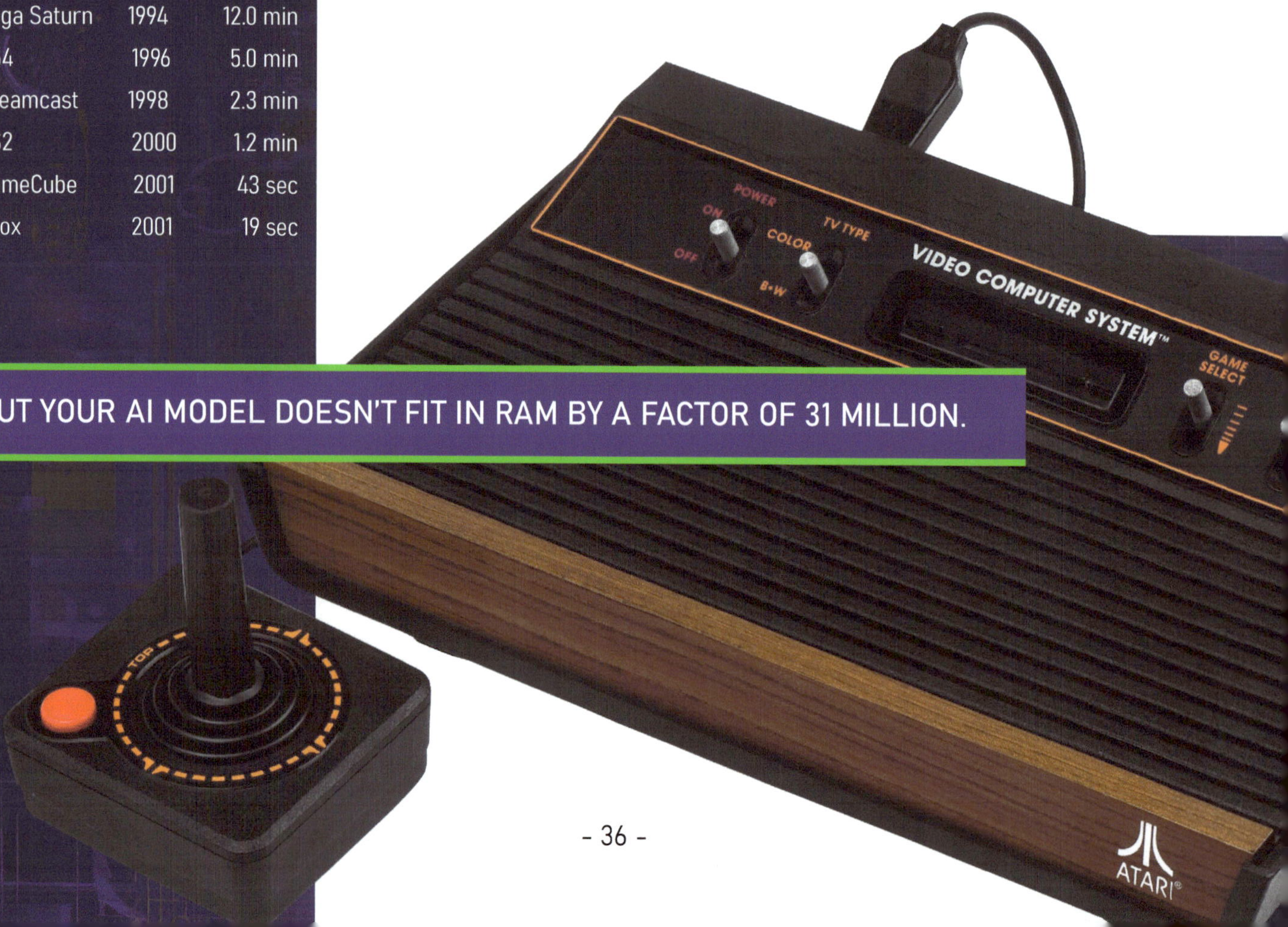

BUT YOUR AI MODEL DOESN'T FIT IN RAM BY A FACTOR OF 31 MILLION.

Run AI on Game Boys

similar brain to graphing calculators, better games

1989-2003
game in a pocket

The original Game Boy runs a Sharp LR35902, a Z80 derivative clocked at 4.19 MHz with 8 kilobytes of RAM. Same processor family as the TI-82 graphing calculator, which runs at 6 MHz. The calculator is faster. Texas Instruments sold a more powerful chip than Nintendo's handheld and used it to graph parabolas. The Game Boy achieves roughly 698,000 multiply-accumulate operations per second and says "Hello" in 5.6 hours. Four AA batteries last 15 hours. Tetris took three minutes to become addictive. Inference takes 6 hours. The Game Boy sold 118 million units. The TI-82 sold to people who had no choice.

THE CALCULATOR IS FASTER. TETRIS IS BETTER.

Can Atari Run AI?

With what: `MOS 6507 @ 1.19 MHz`
Output: `8-bit integer`
Unit: `operations per second`
Compute: `~149,000 ops/sec`

1.1 days. If it survives.

The 6507 has no multiply instruction. Each operation to power AI requires a software loop of shifts and additions, 8 cycles minimum. The RAM situation is beyond absurd, 128 bytes is less memory than this sentence.

So 1 Atari = 100,000x Potatoes

and 1 cyclist = 10 Ataris

What about Game Boy?

With what: `Sharp LR35902 (Z80) @ 4.19 MHz`
Output: `8-bit integer`
Unit: `operations per second`
Compute: `~698,000 ops/sec`

5.6 hours, bring batteries.

Same Z80 family as the TI-82, same software multiply problem. Nintendo clocked it lower to save battery life. Both devices ask an 8-bit processor from 1976 to do work designed for hardware that wouldn't exist for another 40 years.

So 1 Game Boy = 7,000x Potatoes

and 1 cyclist = 143 Game Boys

1964-1980s
literally buried in paperwork

Run AI on Punch Card Computers

AI running on 61 metric tons of paper

The punch card computer was the first time humanity outsourced thinking to a machine and then waited. The IBM System/360 read stacks of paper cards, each one holding a single line of code as rectangular holes. Feed the deck, press start, wait. If you made a mistake, you pulled the card out, repunched it, and reinserted it in the right order. If you dropped the deck, you started over. There was no undo, no screen, no mercy. NASA used these machines to calculate orbital trajectories. Banks used them to process payroll. A skilled operator could process about a thousand cards per minute, and the machine performed roughly 10,000 operations per second. At that rate, saying "Hello" takes about 16 days. Not fast, but the first time a single machine could finish the job in less than a human lifetime.

Can it Run AI?

With what: `IBM 2540 + System/360 Model 30`

Output: `photoelectric card reads + scientific compute`

Unit: `operations per second`

Compute: `~10,000 ops/sec (1.3 KB/sec, 1k cards/min)`

Sure, in 16 days. If you can find 17.5 million cards and a BIG warehouse.

One mainframe, ten thousand operations per second. The math checks out, it just takes 16 days & 17.5M cards.

So 1 punch card = 115,000,000x Potatoes

and 1 cyclist = 1/115th of a punch card mainframe

(and 115 cyclists makes for 1 very sweaty computer room)

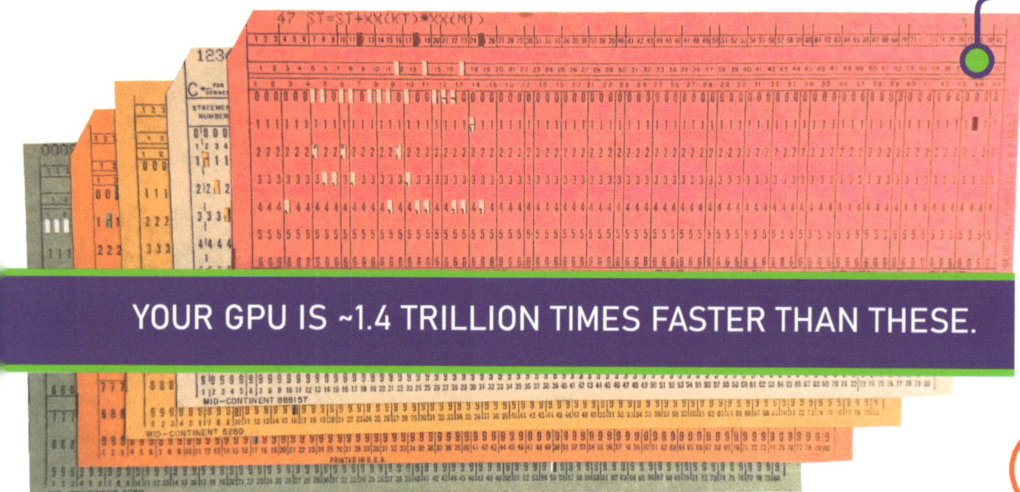

YOUR GPU IS ~1.4 TRILLION TIMES FASTER THAN THESE.

Paper Punch Cards

Mainframes ran Fortran (derived from "Formula Translation"), a high-performance imperative programming language by IBM in 1957. One FORTRAN punch card holds 1 line of code (80 columns, 80 bytes, as rectangular holes in card stock). A programmer in 1965 often used 2,000 cards in a week. ***A quantized 7B AI model would require ~8,750 card boxes at 2,000 cards per box.*** *Don't drop the deck, or you start over.*

NO UNDO. NO BACK BUTTON.

Run AI on Telegraphs

it runs on what hath God wrought, eventually

1844 to present
global calls before Zoom

On May 24, 1844, Samuel Morse sent "What hath God wrought" from Washington to Baltimore in about four seconds. The telegraph consumed 1.5 watts, less than a modern LED bulb, and ran on Grove cells refilled with nitric acid every forty minutes. A professional operator transmits five characters per second. The key closes a circuit, current flows through an electromagnet, an armature strikes a resonator, sound emerges. At fifty percent duty cycle the effective draw is 0.75 watts. Treating each character as one operation, the telegraph says "Hello" in 89 years. The first message traveled 40 miles in 4 seconds. This one travels nowhere in nearly a century. Morse code is not slow because it lacks power. It is as slow as the human wrist.

THE VICTORIAN TELECONFERENCE RAN ON NITRIC ACID AND PATIENCE, LOTS OF IT.

"What hath God wrought" (1844)

1st message in Morse code was 21 characters and took 4 seconds

YOUR AI SAYS "HELLO"

in only 14 billion clicks taking about 89 years

The Telegraph Forever Altered Communication

A telegraph is a device or system for transmitting messages over long distances by converting them into electrical signals, which are sent over wires and decoded at the destination. Developed in the mid-19th century, it revolutionized communication by allowing near-instantaneous, text-based messaging (often using Morse code) over vast distances

The first telegraph message was 21 characters long and changed the world in four seconds. Saying "Hello" through inference takes 700 million times longer. The wrist didn't get slower. The ask got bigger.

Can it Run AI?

With what: Telegraph key+sounder
Output: electromagnetic clicks
Unit: Morse characters/sec
Compute: 5 operations/sec (20 WPM)

89 years. Pack nitric acid.

The telegraph draws 0.75 watts, one of the lowest in this book. The constraint is one wrist, one key, five characters per second, & 14 billion operations between you and "Hello."

So 1 telegraph = 7,500x Potatoes

and 1 cyclist = 130 Telegraph operators

HISTORICAL TOOLS

Ancient engineering that almost got there, if only they had AI.

The power was never the problem. Humans have been generating enough energy to run AI for over a thousand years. They just didn't have anything to plug it into.

The Persians built vertical-axis windmills in the 9th century. The wind mills in Iran still turn. The Dutch scaled the idea up, drained a sea, and accidentally built a 2.4-megawatt wind farm three centuries before the term existed. Tibetan monks spun prayer wheels and may have built the world's most patient distributed computing network. The Romans put workers on treadmills and lifted stones that are still standing. And the churchbell network across Europe was effectively a power grid.

The energy was there. The engineering was extraordinary. It all stopped just short of useful computation, because the missing piece was never power. It was the machine that could do something with it.

Run AI on Roman Treadmill Cranes

100 BC to 1500s
hamster wheels of empires

one person walking lazily could do this

The Romans lifted 300-ton blocks using wooden wheels, rope, and men walking inside drums like hamsters of the empire. The magna rota converted footsteps into vertical force with brutal efficiency, often using 5 workers in a 4-meter wheel generating 500 watts continuously. That is enough to lift 6,000 kg. That's 100 watts per worker, or about the same as riding a bike, just slower, louder, and more likely to involve a centurion yelling at you. One worker walking lazily exceeds the 30-watt threshold needed for AI to say "hello". Some historians say that the Colosseum used roughly 80 cranes during its construction. That's 40,000 watts. Rome could have run an AI deployment while building the arena. They just didn't know it.

ONE ROMAN WALKS IN A WHEEL. THAT'S THE WHOLE THING.

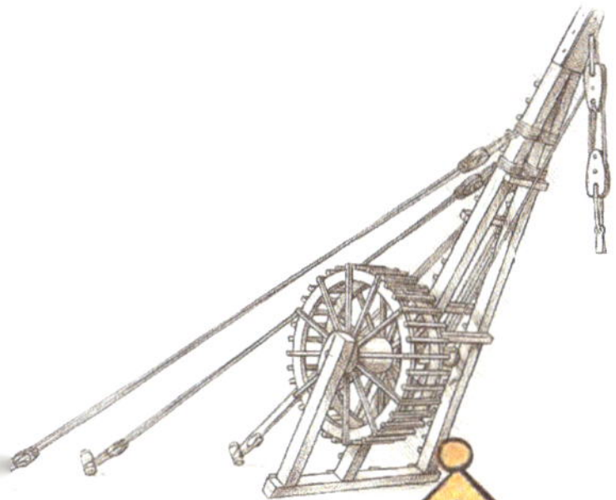

A 13th-century manuscript illustration of a treadmill crane, resembling the working replicas used at the modern Guédelon Castle project in France.

Can it Run AI?

With what: human powered treadmill

Output: rotational power

Unit: workers (walking in wheel)

Watts: ~100W per worker

Yes. One lazy Roman.

One person walking at 0.5 m/s in the wheel produces 100 watts, more than three times the inference threshold. A full 5-person crew generates 500 watts, enough for 16 simultaneous conversations. Walking in circles for AI is historically precedented.

So 1 walker has the power of...

1,000,000x Potatoes

or 1 cyclist = 1

1450-1990s
10,000 of these still standing

Run AI on Dutch Windmills

power that sits scenically in the tulip fields

The Netherlands built over 10,000 windmills between the 15th and 19th centuries, not for aesthetics, though they achieved that accidentally, but for raw mechanical sovereignty. They ground grain, sawed timber, pumped water from polders, and crushed seeds for oil, all without permission from any power grid or fuel supplier. Four fabric-covered sails spanning 24-29 meters catch wind and rotate at 10–20 RPM, driving millstones and pumps through wooden gearing at 20–30% efficiency. They have a sustained output of about 4,000 watts. Historians estimate that at its peak in the 1700s, the Zaan region alone once operated up to 600 mills. That's a 2.4-megawatt wind farm, 3 centuries before the term existed.

ZERO VENDOR LOCK-IN. DOUBLES AS A TOURIST ATTRACTION.

Can it Run AI?

With what: wind-driven sail

Output: rotational shaft

Unit: windmills (sustained ops)

Watts: ~4,000W (avg)

Yes, 133 times over.

One Dutch windmill exceeds the 30-watt threshold by a factor of 133. No batteries, no inverters, no thermal throttling. It turns when the wind blows and requires no explanation to regulators.

So 1 windmill has the power of...

40,000,000x Potatoes

and 1 windmill has the power of...

40 Cyclists

Powering the Dutch Golden Age

Windmills powered the Dutch Golden Age. By the 17th century, over 10,000 mills sawed wood for ships 30 times faster than by hand, pumped water to reclaim land from sea. During WWII, the Dutch resistance used windmill sails to send messages.

Google's Windmill Prank

In 2017 on April Fools Day, Google claimed to have connected 1,170 traditional windmills across Holland to the Google Cloud Platform to control disperse rain-producing clouds with machine learning.

Built in 1994 to commemorate Dutch-Japanese friendship, historic "De Liefde" (meaning "The Love") windmill in Sakura Furusato Square in Sakura City, Japan boasts over 600,000 tulips.

Run AI on Persian Windmills

producing power before the Dutch even started

500 CE to present
1,500 years strong & still counting

The Persian windmill predates the European design by six centuries. Built in what is now Iran and Afghanistan, these vertical-axis machines catch wind through openings in mud-brick walls, rotating like a carousel. No need to face into the wind, no yaw mechanism, no complexity. But the mills in Nashtifan have turned continuously since the medieval period, producing 7,000 watts in steady wind. That's nearly twice the Dutch mill, from a design six centuries older and three times less efficient. The walls channel wind into a brutal funnel. What the design lacks in elegance it compensates with stubbornness. One Persian mill powers 233 conversations.

Persion vs Panemone Windmill

The Persian windmill is similar but distinct from the Panemone windmill. A panemone (from Greek pan + anemos, meaning "all winds") accepts wind from any direction without shielding. The Persian design uses mud-brick walls to block one side and funnel wind to the other, trading omnidirectionality for force.

NOT OLD AT 6 CENTURIES, JUST CLASSIC.

2x the Power Output

With what: wind-driven vanes
Output: rotational shaft
Unit: windmills (8 m/s wind)
Watts: ~7,000W average

Yes, 233 times over.

One Persian windmill in steady wind generates 7,000 watts, enough to run 233 copies of a 7B AI simultaneously, nearly 2x Dutch windmills. And they've been doing it since 500 CE. Nobody asked them to run inference. They didn't need permission.

So 1 windmill has the power of...

70,000,000x Potatoes

and 1 windmill has the power of...

70 Cyclists

DUTCH MILLS WIN ON USING WHAT THEY CATCH. BUT PERSIAN MILLS WIN OVERALL BY CATCHING MORE.

The Dutch mill captures 20-30% of the energy in the wind hitting its sails. The Persian mill only captures 5-15%. But the Persian walls funnel a much larger volume of wind through the vanes. So 10% of a huge amount of wind beats 25% of a smaller amount.

PERSIAN WINDMILL

800 CE
to present day rituals

Run AI on Prayer Wheels

generate power while computing mantras

The Tibetan prayer wheel contains sacred mantras written thousands of times on scrolls wound around a central axis. Each rotation is believed to release every mantra simultaneously, a mechanical broadcast of intention. Some are turned by hand, others by water, wind, or the heat of butter lamps. The rotation is continuous, deliberate, and has been maintained for over a thousand years. We propose using them for inference, which is either a profound insight about computation or a profound insult to a contemplative tradition. Possibly both.

WHEEL TURNS. MANTRAS MAKE POWER.

The Prayer Wheel is Hand-Cranked with Sacred Mantras

A hand-cranked prayer wheel produces about 1 watt per rotation per second. Water-powered wheels achieve 3-5 watts. Large temple wheels reach 10. But here's the hybrid question: a typical scroll contains 100,000 repetitions of a mantra, all "processed" simultaneously with each turn. If you count each mantra as one operation, the prayer wheel performs 100,000 operations per second. That's more than a slide rule, an astrolabe, or a pinwheel calculator. By mantra count, it's the most powerful analog computer in Section 2. It just computes prayers instead of matrix multiplications.

Can it Run AI?

With what: `hand-cranking`

Output: `rotational energy`

Unit: `prayer wheels (1 rotation/sec)`

Compute: `~100,000 mantras/sec (per scroll)`

30 monks, 30 wheels.

Thirty practitioners turning prayer wheels in unison produce 30 watts, enough to say "Hello" in about 6 seconds.

So 1 wheel produces the power of about...

10,000x
Potatoes

and 1 cyclist's power takes about...

100
Prayer Wheels

Run AI on Monastic Bells

ask not for which AI model the bell tolls

500-1100 AD

a powerful divine interruption

The Benedictine Rule divided the day into 8 canonical hours, each announced by bell. Vigils at 3am. Lauds at dawn. Compline at 9pm. For a thousand years, these bells formed one of Europe's earliest distributed timekeeping networks. A monastery bell weighing 500 kilograms, swung through a 90-degree arc, delivers approximately 2,000 watts of mechanical power for the three seconds it takes to complete one ring.

When a monk pulls the rope, human muscle power (roughly 300 watts during the pull) transfers through a wheel mechanism to swing several hundred kilograms of bronze. The energy releases in a burst, part kinetic, part acoustic, mostly dissipated as heat in the bell metal and air resistance. The sound itself radiates at 100-120 decibels, carrying about 1 watt of acoustic power. But sound is not the substrate. The mechanical work is.

The bell rings 40 to 60 times per day. Each ring is 3 seconds of power, then silence until the next hour. Averaged over 24 hours, those 60 rings at 2,000 watts each deliver a continuous power of approximately 4.2 watts. More than 40,000 potatoes, delivered by rope.

With what:
~500kg bronze bell

Output:
Mechanical rope pulls

Unit:
Canonical bells (60x rings per day)

Watts:
~4.2W averaged over 24 hours

That means 1 bell has the power of...

42,000x Potatoes

and 1 cyclist =

24 Bells

Can it Run AI?

Medieval Europe had roughly 40,000 monasteries. If every bell rang simultaneously and continuously, that's 167,000 watts, enough for 5,500 saying "hello" from your AI. But bell rings are as hard to accumulate as prayers.

Yes, but only if we link 40,000 monasteries somehow.

MONKS ACCIDENTALLY BUILT A 167 kW GRID. WE CAN'T KEEP MICROSERVICES ONLINE.

Part 2

BIOLOGICAL COMPUTE

Section 5

Animals & Pets

It's warm, fuzzy (or feathered) and makes power to run AI.

Every animal on Earth generates power. Most of it is useless. Cats produce static electricity when you pet them. Hamsters spin wheels for no strategic reason. Pigeons carry data faster than 5G but cannot do math. Salmon swim upstream at 100 watts per fish, which is exactly one cyclist with much worse life planning. Elephants shake the ground. Whales shake the ocean. Old MacDonald had a farm, and kilowatt-makers walk around eating corn and grass.

These are your pets, your livestock, and the animals on your screensavers. None of them were designed to run AI. The chicken has never once, in the entire history of chickens, volunteered for anything. But every one of them converts food into power. The math does not care whether that power comes from a data center or a cow.

Run AI on Cat Static

you'd need a cat-astrophic amount of kittens

38 chromosomes
ZAP! OWW! My whiskers!

The United States has approximately 76.3 million cats per the American Veterinary Medical Association. A cat generates static electricity when you pet it, and the triboelectric effect transfers electrons between fur and skin and makes a tiny spark with discharge voltage about 5,000 volts, which sounds terrifying until you learn the current is negligible. One cat produces about 1 milliwatt. That's 10 potatoes and absolutely no AI. But scale changes everything. If every cat in America were being petted at the same time, the combined output would be 76,300 watts, enough for AI to say "hello" about 2,500 times simultaneously. The cats do not care. They were going to be petted anyway.

AI IS TRANSLATING MEOWS

AI tools like the **MeowTalk** *App, are >90% accurate at translating cat vocalizations.*

PURR-HAPS WE PET 30,000 CATS. FOR SCIENCE.

Table of Cat Power, with All the Cats

Scale	*Cats*	*Power*	*AI Hello's*
Your cat	*1*	*0.001W*	*not happening*
A cat cafe	*20*	*0.02W*	*Still No*
Neighborhood	*50*	*0.05W*	*No*
All US cats	*76,300,000*	*76,300W*	*2,500*
Cats worldwide	*~600,000,000*	*600,000W*	*20,000*
All cats ever born	*~10 billion(est)*	*10,000,000W*	*irrelevant*

~5,000 Volts
(terrifying, harmless)

If all the world's cats were petted at once it would create enough power for AI to say "hello" about 20,000 times. Not a viable power source, but the best afternoon imaginable.

Can it Run AI?

With what: triboelectric fur discharge
Output: static electricity
Unit: cats (actively petted)
Watts: ~0.001W per cat

Yes, 30,000 cats. Petted. At once.

One cat produces 1 milliwatt. You need 30,000 of them. The voltage is impressive. The current is not. This is the most on-brand chapter in the book and the least useful.

So 1 cat = 10x Potatoes

and 1 cyclist = 100,000 Cats

(please do not attempt this)

44 chromosomes
so so close to human 46

Syrian hamsters (Mesocricetus auratus) typically have 44 chromosomes, while Chinese hamsters (Cricetulus griseus) have a diploid number of 22 chromosomes. Fun fact, Chinese Hamster Ovary (CHO) cells are important in modern research.

Run AI on Hamster Wheels

60 hamsters, running in the dark

There are many different hamster species. The common pet hamster, the syrian hamster runs 5 to 8 kilometers (or 3 to 5 miles) per night on its wheel with the unhinged dedication of an organism that has no concept of futility. Attach a micro-generator and the wheel produces about 0.5 watts during active running. That's more than a cat by a factor of 500, which the hamster would find satisfying if hamsters understood ratios. Sixty hamsters running simultaneously produce 30 watts, enough for inference. The problem is timing: hamsters run 4 to 6 hours per night and sleep all day, giving you roughly 20% duty cycle.

Each hamster in a tiny vest makes about 0.1 to 0.15 volts

(running in wheels is more efficient)

Wearing tiny electric harvesting vests

In 2009, Georgia Tech researchers fitted dwarf hamsters with nanogenerator vests lined with zinc oxide nanowires and let them run. Each hamster generated approximately 60 microwatts, enough to measure, not enough to matter. The team calculated it would take over 20 million vested hamsters to power a single American household.

To get AI to say "hello" one time, you need 500,000 hamsters wearing matching vests or 60 tireless hamsters running in wheels. The hamsters may have opinions on the dress code. The original hamster-vest researchers were funded by DARPA, which presumably had grander ambitions for the technology than this.

THEY RUN AT NIGHT. SCHEDULE ACCORDINGLY.

Six month old Winter White Russian Dwarf Hamster. Not presently an AI user. Maybe

Hamster Electric Wire Chewing Hazards

Hamsters have constantly growing teeth and are prone to chewing through household electrical cords, requiring owners to secure all wiring.

HAMSTERS OUTSMARTING AI AT MAZES

Hamsters (along with rats and mice) are evolutionarily gifted at navigating mazes, often outsmarting traditional AI maze solvers.

Adult Roborovski hamster

Can it Run AI?

With what: hamster exercise wheel

Output: mechanical to electrical

Unit: hamsters (running)

Watts: ~0.5W per hamster (active)

Yes, 60 hamsters. At night.

Sixty hamsters running simultaneously produce 30 watts. But they sleep all day, so for 24-hour coverage you need 273 in rotating shifts. The schedule is the bottleneck, not the hamsters. They'll run regardless.

So 1 hamster makes the power of...

5,000x
Potatoes

and 1 cyclist's power takes about...

200
Hamsters

HAMSTERS CAN RUN 3 TO 6 MILES PER HOUR (5-9.6 KM/H), RUNNING SEVERAL MILES IN A SINGLE NIGHT.

80 chromosomes
and 1,500 to 2,500 feathers

Can't Run AI on Carrier Pigeons

but you can move the weights across continents

A carrier pigeon flies at 60 miles per hour with a terabyte microSD card strapped to its leg. Over 50 miles, that's 50 minutes of flight time and an effective throughput of 350 megabytes per second, which beats 5G on raw payload over short distances. Over 1 mile the pigeon achieves 17 gigabytes per second, faster than 10-gigabit fiber. The bandwidth is extraordinary, but latency still has issues.

The pigeon carries roughly 75 grams of payload. Stuffed with SD cards, your bandwidth is god-tier.

PIGEON LATENCY IS TERRIBLE.

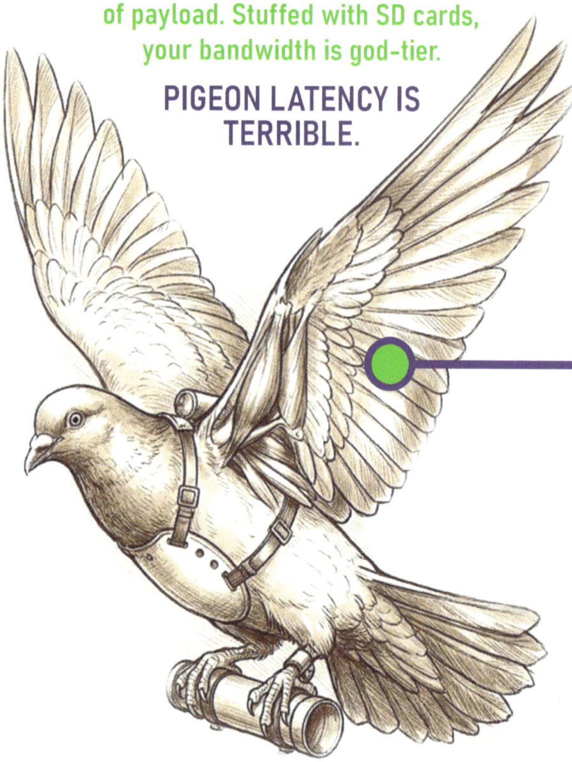

Can it Run AI?

With what: `pigeon + microSD card`
Output: `data transfer(not compute)`
Unit: `pigeon-flights`
Throughput: `~350 MB/s(50 miles)`

No, it can't compute.
But it can deliver.

The pigeon doesn't run inference. It carries the weights. 1T on a microSD, 50 miles, 50 minutes. This is a logistics chapter, not a compute chapter, and the distinction matters.

So 1 pigeon's flight power...
100,000x
Potatoes
and 1 cyclist's power takes about...
10
Pigeons

Distance	Time	Speed
1 mile	1 min	17 GB/s
10 mile	10 min	1.7 GB/s
50 mile	50 min	350 MB/s
100 mile	100 min	175 MB/s

For comparison: 4G = 6 MB/s, 5G = 125 MB/s. The pigeon beats 5G at any distance under 150 miles.

Brussels erected a monument to Pigeon-Soldat.

Pigeons cannot run inference, but can deliver the model weights

RFC 1149, "A Standard for the Transmission of IP Datagrams on Avian Carriers," was published on April 1, 1990, and actually tested in Bergen, Norway in 2001. Nine pigeons were sent. Four arrived. Packet loss: 55%. Seven gigabytes on a microSD, 50 miles in 50 minutes. By the time your 4G downloads a model, the pigeon is already home. This is the only chapter in the book about data transfer instead of power, because sometimes the bottleneck isn't computation. It's getting the weights to the computer.

World War I German Camera Pigeon (right) and British Soldier in France Holding a Carrier Pigeon (National Library of Scotland)

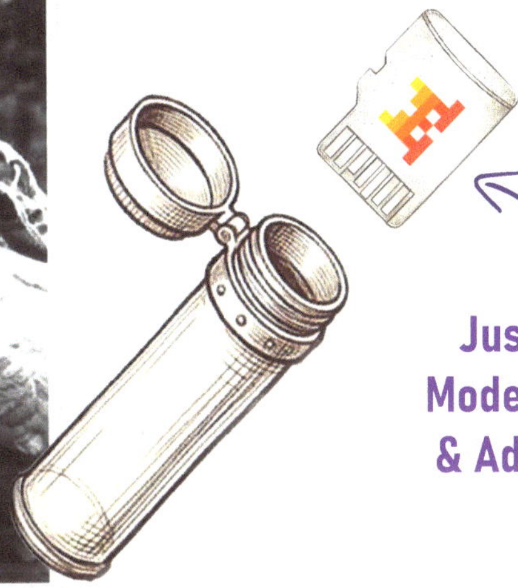

Just Insert Model Weights & Add Pigeon

The Brave Pigeons of War

During World War I, there was no text or email. So both sides deployed homing pigeons to carry messages across front lines. Birds flew through artillery fire and gunfire. More than 100,000 pigeons were deployed and some even won awards for their valor, including the French Croix de Guerre and the Dickin Medal, the animal equivalent of the Victoria Cross.

55% PACKET LOSS. BETTER THAN WIFI.

Champion Pigeon Sold for €1.6 Million

Carrier pigeons average 90 km/h over distances up to 1,800 kilometers. Racing pigeons have been clocked at 177 km/h with tailwinds. Belgium alone has over 18,000 active pigeon racers and a champion bird sold for €1.6 million in 2020.

SPECIALIZED TRUCK FOR TRANSPORTING RACING PIGEONS.

44 chromosomes
weirdly equal to a hamster

Run AI on Whale Songs

krilling me softly with your song

Whales make some of the loudest sounds of any animal. The sperm whale is the loudest on Earth at 230 decibels (or higher) for a short click lasting about ~100 microseconds. Quieter than the sperm whale but still louder than a jet, the blue whale call can hold a sustained sound at 188 decibels that can travel up to 1,000 miles through the ocean. The acoustic power it radiates is approximately 2 watts. That's 40,000 times louder than a human voice, which produces about 50 microwatts of acoustic energy during normal conversation. You would need 15 whales singing simultaneously to reach 30 watts of acoustic power. The metabolic cost to produce the sound is higher (~75W), meaning half a whale's vocalization could power AI. But only about 2W of that actually propagates through the ocean to be feasibly harvested. As a power source, the loudest sustained animal sound on Earth makes less usable power than a USB charger.

Can it Run AI?

With what: acoustic energy
Output: sound waves (10-40 Hz)
Unit: singing blue whales
Watts: ~2W acoustic per whale

Yes, 15 singing whales.

Fifteen blue whales singing simultaneously produce 30 watts of acoustic power. The sound travels 1,000 miles. The power doesn't. You'd have to be very close to a very large animal.

So 1 whale = 20,000x Potatoes

& 1 cyclist = 50 Whales (acoustically)

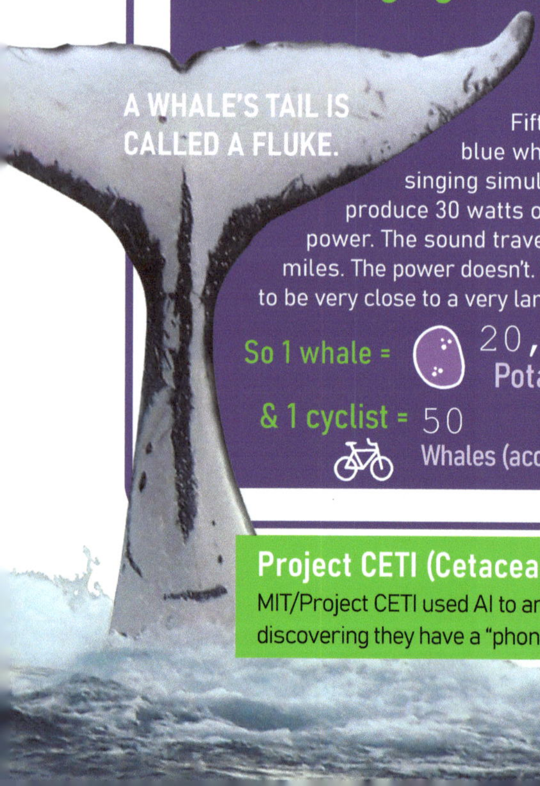

A WHALE'S TAIL IS CALLED A FLUKE.

LOUDEST ANIMAL ON EARTH.

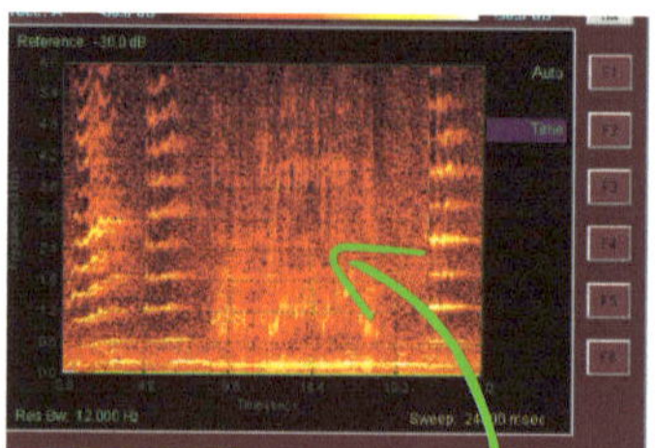

A **SPECTROGRAM** *of a whale song is a visual, graph-based representation of underwater sound, plotting frequency over time to reveal complex, structured vocalizations.*

ML, MUSIC, & MEANING

Machine learning identified information in specific in timing and types of musical-like expressions (RUBATO) & added sounds (ORNAMENTATION) in whalesong.

Project CETI (Cetacean Translation Initiative)

MIT/Project CETI used AI to analyze of sperm whale clicks, discovering they have a "phonetic alphabet."

Run AI on Elephant Footprints

6,000 kilograms on the ground every second

An African elephant weighs 6,000 kilograms and walks at roughly one step per second. Each footfall sends Rayleigh waves through the earth detectable more than 10 miles away, letting them communicate seismically through their feet. The total seismic energy per walking elephant is about 3 watts, most of which dissipates as heat in the ground. If you paved their path with piezoelectric tiles, you could harvest roughly 200 milliwatts per step. You would need 150 elephants walking continuously to reach 30 watts via piezo harvesting, or 10 elephants if you could capture the full seismic output. The universe gave them earthquake feet and cancer resistance. It did not give them GPUs.

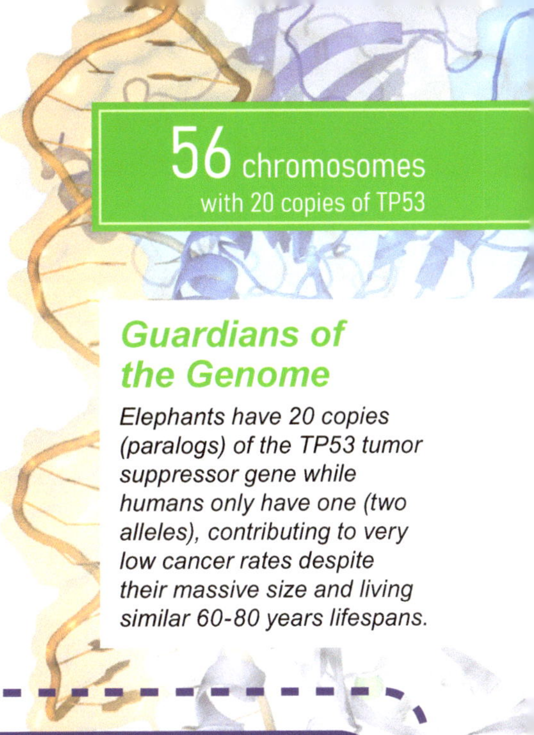

56 chromosomes
with 20 copies of TP53

Guardians of the Genome

Elephants have 20 copies (paralogs) of the TP53 tumor suppressor gene while humans only have one (two alleles), contributing to very low cancer rates despite their massive size and living similar 60-80 years lifespans.

Source	Decibels	Acoustic Watts
Whisper	30 dB	0.000000001W
Talking	60 dB	0.00005W
Rock concert	110 dB	10.0W
ELEPHANT STEP	117 dB	3.0W
Fighter jet	140 dB	1,000W
BLUE WHALE	188 dB	2.0W
SPERM WHALE	230 dB	~10,000W

NOTE: WATER CHANGES SOUND

Approximate source power comparison. Whale decibels are measured in water using a 1 micropascal reference. Air decibels use a 20 micropascal reference. This makes underwater numbers roughly 26 dB higher for the same pressure.

Can it Run AI?

With what: piezoelectric ground
Output: seismic to electrical
Unit: elephants (walking)
Watts: ~0.2W harvestable per elephant

Yes, with 150 elephants walking on special tiles.

The seismic communication is more impressive than the power output. Elephants solved long-range networking through the ground. The 200mW per step is a footnote to a much better engineering achievement.

So 1 elephant = 2,000x Potatoes

and 1 cyclist = 500 Elephants

THE GROUND COULD BE A POWER GRID.

80 chromosomes
more chromosomes than you

Swifts Landscape Sketches
1887 drawing by Vincent van Gogh showcasing his shift toward faster, more energetic strokes. Appropriate.

Run AI on Migrating Swifts

83% uptime, zero maintenance windows

The common swift stays airborne night and day for ten months without landing. It sleeps on the wing. It eats, drinks, and mates in flight. It migrates 14,000 miles per year between Europe and sub-Saharan Africa on a 40-gram body burning 1.5 watts of continuous metabolic power.

How Many AI Hello's from a Flock of Swifts?

That's 83% uptime across 7,200 hours of uninterrupted flight, which is better than most cloud providers manage with air-conditioned server rooms and on-call engineers. Twenty swifts in formation produce 30 watts, enough for an AI to say one "hello." A migrating flock of 5,000 generates 7,500 watts, enough for AI to say "hello" 250 times, delivered without downtime across two continents.

AI AND MIGRATORY BIRDS

AI is being used to help track and analyze bird migration, for example MistNet is an AI developed by University of Massachusetts Amherst and the Cornell Lab of Ornithology using vast archives of weather radar data

Can it Run AI?

With what: `flight power`
Output: `aerobic watts`
Unit: `swifts`
Watts: `~1.5W per bird`

Yes, 20 swifts.

Each swift burns 1.5 watts, making 30 watts for an AI to say "hello"

So 1 swift = 15,000x Potatoes

& 1 cyclist = 67 Swifts

(but the swifts have better uptime)

Salmon produces 67x more power per fish.
Swifts produce power for 67x longer.

Migration optimizes for either burst or endurance. Nature picked endurance. Data centers picked burst. The salmon would like a word.

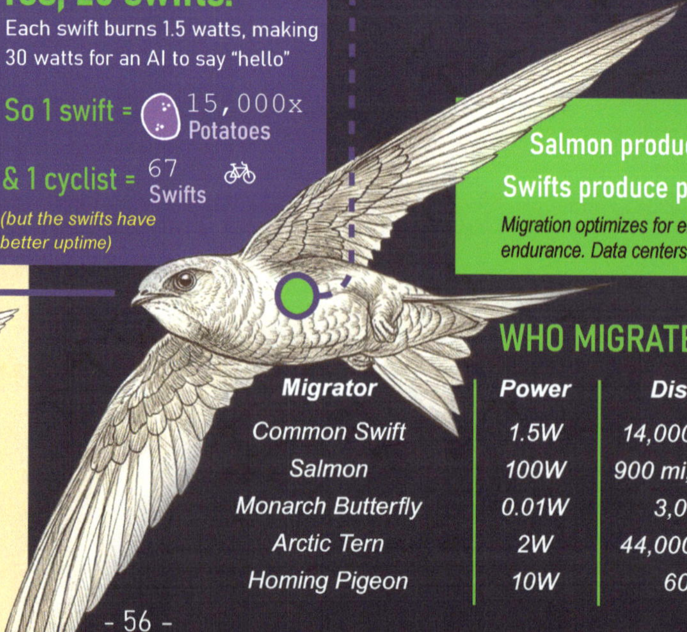

WHO MIGRATES BEST?

Migrator	Power	Distance	Survives?
Common Swift	1.5W	14,000 mi/year	Yes, for 20+ years
Salmon	100W	900 mi, one way	No
Monarch Butterfly	0.01W	3,000 mi	Technically (multi-generational)
Arctic Tern	2W	44,000 mi/year	Yes
Homing Pigeon	10W	600 mi	Yes (with a microSD card)

Run AI on Migrating Salmon

100 watts, 900 miles, one way

74 chromosomes
DNA that takes a long hike

A Chinook salmon swimming upstream during spawning migration produces roughly 100 watts of sustained metabolic power. That is exactly one cyclist. The fish weighs 15 kilograms, swims at 1 to 2 meters per second against the current, covers up to 900 miles, stops eating for the entire journey, and dies at the end. One salmon exceeds the 30-watt inference threshold by a factor of three. A single fish could power three AIs saying "hello" simultaneously for the duration of its migration, which lasts weeks to months depending on distance. A spawning run of 100,000 fish generates 10 megawatts, enough AI to say "hello" 333,333 times. That means that salmon power could run full AI conversations. It is swimming upstream to reproduce and die. The power is a byproduct of desperation, not design.

SALMON MAKE GREAT PROGRAMMERS. THEY ARE ALWAYS STREAMING.

Salmon are remarkable fish that travel thousands of miles from the ocean to the exact freshwater stream where they were born to spawn, often leaping up to two meters over obstacles

Can it Run AI?

With what: metabolic swimming power
Output: sustained aerobic watts
Unit: migrating salmon
Watts: ~100W per fish

Yes. One fish. It won't last.

One salmon produces 100 watts, more than three times the inference threshold. The power is real. The duration is limited. The fish has other priorities and will be dead before you finish fine-tuning.

So 1 salmon = 1,000,000x Potatoes

and 1 cyclist = 1 Salmon (identical wattage, different survival rate)

VARIES
by what you keep on the farm

Run AI on the Farm

one horse is one horsepower, not a coincidence

A horse produces one horsepower. This is not a metaphor or a rough estimate. James Watt measured the sustained pulling power of draft horses in 1782 and defined the unit accordingly. One horsepower equals 746 watts, enough for 25 simultaneous AI conversations. The horse is the only organism in this book where the animal is a common unit of measurement.

Running on Cow Power

A dairy cow contributes differently. Body heat alone accounts for 500 watts from a 600-kilogram ruminant. But the methane is where it gets interesting: a cow produces 250 to 500 liters of methane per day. Burned in a generator at 30% efficiency, that's 38 watts continuous. In a fuel cell at 50%, it's 62 watts. One cow's burps-and-farts (methane), captured and converted, could produce more power than 60 hamsters running all night. The United States has 9.4 million dairy cows producing a combined 352 megawatts of capturable methane power, or roughly 12 million conversations.

The Barn Table

Animal		Watts	Source	For 30W	AI Hello's
Horse	64	746W	mechanical pull	0.04	25
Draft Ox	60	500W	mechanical pull	0.06	17
Dairy Cow (heat)	60	500W	body heat	0.06	17
Dairy Cow (methane)	60	38-62W	captured methane	0.5-0.8	1-2
Pig	38	150W	body heat	0.2	5
Goat	60	75W	body heat	0.4	2.5
Sheep	54	70W	body heat	0.43	2.3
Chicken	78	10W	body heat	3	0.3

The pig has 38 chromosomes. The cat has 38 chromosomes. The pig produces 150,000 times more power. Chromosomes explain very little about power.

The Rest of the Farm Scales Predictably

A pig produces 150 watts of body heat. A sheep or goat, about 70. A draft ox pulling matches a horse at 500 watts.

An Udder-ly Reasonable Way to Power AI

The Chickens are the Surprise

Each bird generates only 10 watts of body heat, but an industrial chicken house holds 20,000 birds producing 200 kilowatts, enough for 6,667 conversations. The United States has 518 million chickens generating a combined 5.2 gigawatts of body heat, which is more than most countries' nuclear programs. The farm has always been a power plant. We just measured it in milk and eggs instead of watts.

Also not there is an academic parody paper called "Chicken Chicken" where the entire text is just the word "chicken" used to explain complex computer science concepts

AI NOW IDENTIFIES & MONITORS COWS

Ida (Intelligent Dairy Farmer's Assistant) is an AI-powered system by Connecterra that monitors cow health, heat, and behavior in real-time.

OLD McDONALD HAD AN A-I-A-I-O.

Can it Run AI?

With: heat+methane+mechanical
Output: depends on the animal
Unit: farm animals
Best: horse, 746W
Worst: chicken, 10W

One horse. 25 conversations.

But also, one cow's methane (38-62W), five pigs (750W), or 3 chickens per AI "hello," with 20,000 in the barn.

So 1 horse = 7,500,000x Potatoes

The horse beats the cyclist 7.5x.

Farm	Animals	Power	AI Hello's
One horse	1	746W	25
Chicken house	20,000	200 kW	6,667
Dairy farm (200 cows)	200	100 kW (heat)	3,333
All US dairy cows (methane)	9,400,000	352 MW	11,700,000
All US chickens (heat)	518,000,000	5.2 GW	173,000,000

CHICKENS WIN BY VOLUME. THE HORSES WIN PER HEAD. THE COWS WIN BY BEING ON FIRE FROM THE INSIDE.

Section 6

From the Sea

Where fins, claws and flagella make biological power we can harness for AI.

The ocean has been running experiments in electricity for 500 million years, and nobody told the engineers. Torpedo rays built capacitors out of muscle. Electric eels stacked cells in series long before Volta made it respectable. Stargazer fish turned modified eye muscles into weapons. Electric catfish were throwing 350 volts while the Egyptians built pyramids.

But voltage is only half the story. Dinoflagellates turn wave energy into light visible from space. Sonoluminescence makes tiny stars from crashing water. Jellyfish run distributed nerve nets on a microwatt with no brain and no concern for your infrastructure budget. Beneath all of it, a quintillion copepods drift through the dark, each producing almost nothing and all of them together producing more power than entire human systems. The ocean had a 500-million-year head start on R&D.

Run AI on Torpedo Rays

nature's natural capacitor found in the ocean

The Atlantic torpedo ray generates 220 volts at 8 amps through roughly half a million electric plates arranged in about a thousand vertical columns, each cell contributing ~0.15 volts. Peak discharge: 1,760 watts for fifty milliseconds. Then the sodium-potassium pumps spend thirty minutes laboriously rebuilding the charge. Duty cycle: 0.003%. Sustained average output: 49 milliwatts. Romans stood on live torpedo rays to treat gout. Scribonius Largus, physician to Emperor Claudius, prescribed them for headaches in 46 CE. The Latin torpedo means "numbness," which tells you everything about the patient experience and nothing about the sustained power output.

28-86 chromosomes
more DNA than you, less time

Chromosomes in torpedo rays (family Torpedinidae) varies by species

Torpedo marmorata (Marbled electric):	*2n = 86*
Torpedo tokionis:	*2n = 86*
Torpedo californica:	*2n = 86*
Torpedo torpedo:	*2n = 66*
Torpedo nobiliana:	*2n = 66*
Narcine brasiliensis (Lesser electric):	*2n = 28*

ROMANS STOOD ON THEM FOR HEADACHES.

The Roman physician Scribonius Largus (c. 1–50 AD) described the treatment in his Compositiones Medicae. In the procedure, a live torpedo ray (often Torpedo nobiliana) was placed under a patient's feet until their entire foot and leg up to the knee was numbed by the electric discharge (up to 220 volts) generated by the fish's specialized muscle cells in their pectoral fins. But they never thought of using it for powering anything, especially not AI.

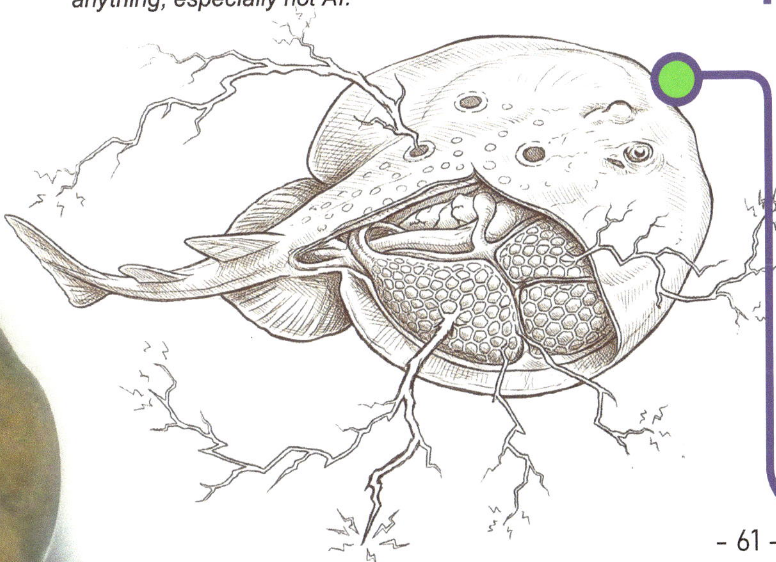

Can it Run AI?

With what: bioelectric capacitor discharge

Output: 220V × 8A pulse

Unit: ray-shocks

Watts: ~0.049W (sustained average)

Yes, with 612 rays taking turns.

Each ray fires for 50ms, then recharges for 30 minutes. The peak is extraordinary. The average is not. This is the duty cycle problem in its purest form.

So 1 torpedo ray = 490x Potatoes

and 1 cyclist = 2,041 Torpedo rays

52 chromosomes
unique genes for electric power

Electrophorus voltai is the strongest bioelectric animal, producing up to 860 volts. This superpower is linked to specialized genes making its muscle cells act like specialized electrocyte "batteries" arranged in series.

When E. voltai releases high-voltage shocks, it sometimes triggers transfer DNA transfer in fish and nearby animals, actually changing DNA in its environment.

Run AI on Electric Eels

the fish that inspired the battery

The electric eel is not an eel. It is a knifefish, and 80% of its body is electric organ. Three separate systems stack electrocytes in series: the main organ for high-voltage stunning, Hunter's organ for medium discharge, and Sachs' organ for electrolocation. Electrophorus voltai, the highland species identified in 2019, produces 860 volts at roughly one amp (peak power 860 watts). Each pulse lasts two milliseconds.

The fish manages about 400 pulses per day during active hunting. Total time spent discharging: 0.8 seconds out of every 24 hours. Duty cycle is about 0.0009%. The most famous electric animal on Earth spends 86,399.2 seconds per day doing nothing electrically interesting.

IT IS MOST ELECTRICAL ANIMAL ON EARTH.

An Electrical Inspiration

The fish that inspired all of electrical engineering produces less continuous power than a phone charger, and the phone charger doesn't need to eat.

Until 2019, scientists believed there was one species of electric eel. ML-assisted DNA analysis revealed 3.

Species	Voltage	Habitat
E. electricus	480V	Guiana Lowland
E. varii	151V	Deep river
E. voltai	860V	Highland rapids

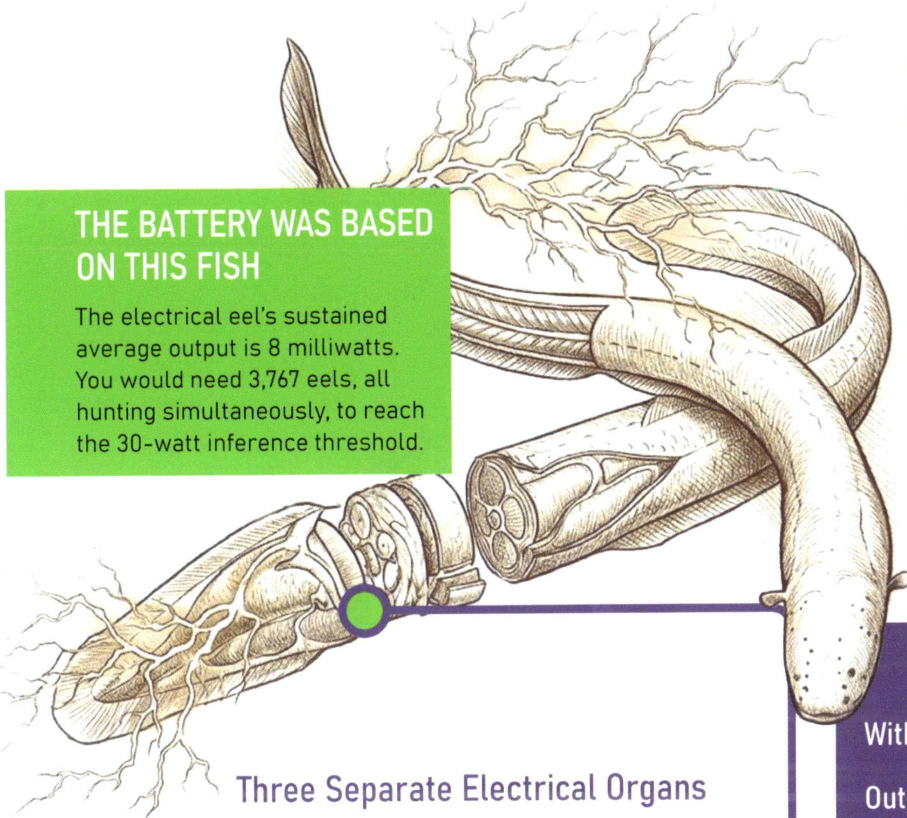

THE BATTERY WAS BASED ON THIS FISH

The electrical eel's sustained average output is 8 milliwatts. You would need 3,767 eels, all hunting simultaneously, to reach the 30-watt inference threshold.

VOLTAIC PILE (also in this book) was modeled on eels

In 1800, Alessandro Volta stacked alternating discs of zinc and copper separated by brine-soaked cloth and produced the first continuous electric current. He called it the "voltaic pile." He modeled it explicitly on the stacked electrocytes of the electric eel. The battery is based on this fish. The unit of electrical potential (the volt) is named after Volta, who studied electric eels. Every battery in every device you have ever owned descends architecturally from an animal that uses its power for under a second per day.

Three Separate Electrical Organs

The eel has three separate electrical systems for three separate jobs. The high-voltage main organ is a weapon. Sachs' organ is a sensor, pulsing continuously at low voltage to map the environment. The eel is simultaneously a taser and a sonar array. Neither function produces sustained power worth harvesting.

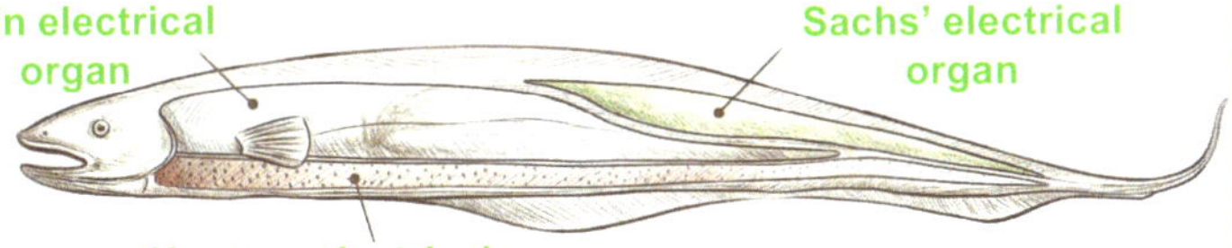

THE THREE ELECTRICAL ORGANS

ORGAN	VOLTAGE	PURPOSE
Main	*600-860V*	*Stunning prey*
Hunter's	*~220V*	*Navigation, low-power pulses*
Sachs'	*~10V*	*Electrolocation (continuous, weak)*

Can it Run AI?

With what: `bioelectric discharge (stacked electrocytes)`

Output: `860V × 1A for 2ms`

Unit: `eel shocks`

Watts: `~0.008W (sustained average)`

Yep, only 3,767 eels needed. Shocking simultaneously.

Each eel fires for 2 milliseconds, 400 times per day. Peak is spectacular. Sustained is 108,000 times worse. This ratio (peak versus sustained) is the entire story of electric fish.

So 1 eel = 80x Potatoes

and 1 cyclist = 12,500 Electric eels

28-48 chromosomes
organs made from eye muscles

Run AI on Stargazer Fish

buried in sand, furious about it

The northern stargazer buries itself in sand with only its eyes and mouth exposed. When something gets close enough, it discharges fifty volts through modified extraocular muscles that evolution repurposed from eye control into biological capacitors. The fish that stares at stars is powered by its own eyeballs (well technically modified muscle tissue just posterior to the eyes, not the eyeballs themselves). Its peak output is 50 watts for 75 milliseconds. But stargazers are ambush predators. They lie motionless for hours. Maybe fifteen discharges per hour when actively hunting, more likely fewer. Sustained average: 16 milliwatts. We first thought of this as 2.5 watts using a 5% duty cycle, but that would require the fish to be discharging three seconds out of every minute. A fish that hunts by lying perfectly still does not fire its weapons 5% of the time. It fires when something is close enough to eat. The corrected number is 160 times worse and substantially more honest.

Can it Run AI?

With what: bioelectric discharge (modified eye muscles)

Output: 50V × ~1A for 75ms

Unit: irritated discharges

Watts: ~0.016W (sustained average)

Sure, with 1,875 furious stargazers.

The least impressive electric fish in this section, but the only one powered by repurposed eyeball muscles, which counts for something.

So 1 stargazer = 160x Potatoes

and 1 cyclist = 6,250 Stargazers

GENERATING 50W WITH EYE MUSCLES IS A FLEX.

Run AI on Electric Catfish

350 volts from the time of the pharaohs

56 chromosomes
catfish genes range 48 to 60[8]

The Nile electric catfish appears in Egyptian tomb paintings from 2750 BCE, shown with radiating lines often interpreted as ancients understanding exactly what it did. Up to 2 million electrocytes stacked in series along its body produce 350 volts at one amp, enough to stun a horse. Each pulse lasts fifteen milliseconds. Then the fish needs ten to fifteen minutes to rebuild its ionic gradients via ATP synthesis. Four maximum discharges per hour before exhaustion. The brain's discharge command nucleus sends simultaneous signals to every electrocyte, creating a coordinated pulse that propagates down the body at ten meters per second. Averaged over recharge time, each catfish generates 8 milliwatts of sustained power, less than a single AA battery but with significantly more voltage drama. The Egyptians drew it on tomb walls 4,700 years ago, we're going to harness it.

Electric Catfish in Ancient Art
A bas-relief at the Mastaba of Ti in Saqqara, Egypt, shows the electric Nile catfish.

OLD WHEN PYRAMIDS WERE NEW.

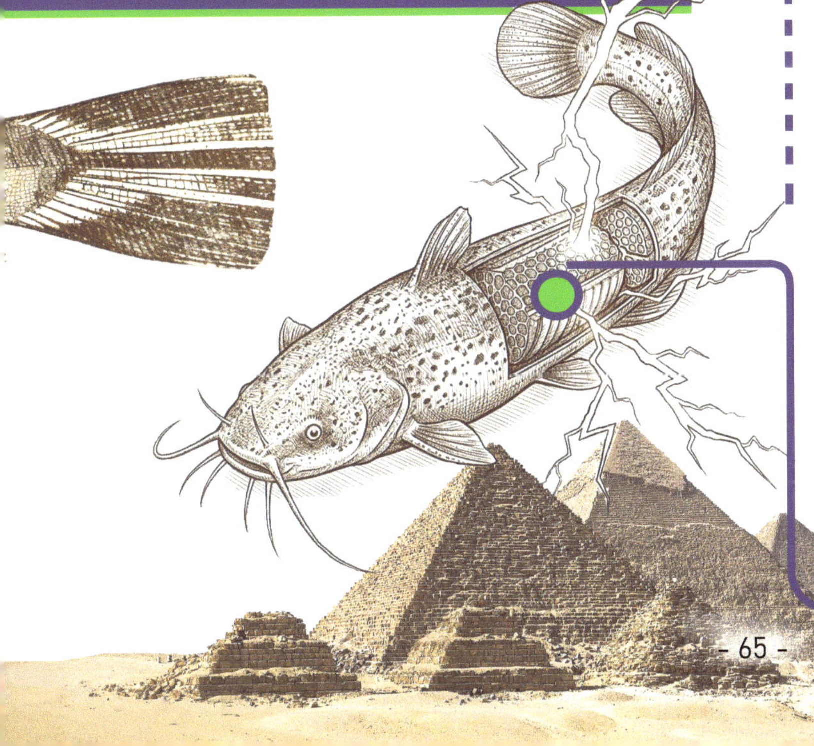

Can it Run AI?

With what: bioelectric pulse (series electrocytes)
Output: 350V × 1A for 15ms
Unit: pharaonic discharges
Watts: ~0.008W (sustained average)

3,750 catfish. Very tired.

Four discharges per hour, fifteen milliseconds each. The voltage is pharaonic. The duty cycle is not.

So 1 electric catfish is the power of about... 80x Potatoes

and 1 cyclist = 12,500 Electric catfish

88 chromosomes
10,400 g acceleration

Run AI on Mantis Shrimp

the fastest strike in the animal kingdom

The mechanism is a saddle-shaped spring in the shrimp's raptorial appendage. Muscles slowly load elastic energy over two to three seconds, then release it catastrophically in 3 milliseconds. The strike delivers approximately 1 joule of kinetic energy per punch. That's comparable to acceleration of a .22 caliber bullet. The cavitation bubble that forms in the strike's wake collapses asymmetrically, generating a secondary shockwave, plasma-temperature compression, and a flash of light from raw mechanical force. The shrimp sees the flash. It has sixteen types of photoreceptor cells, compared to your three. It sees colors you cannot name in wavelengths you cannot detect, and it punches light into existence as a side effect of lunch.

Sustained output depends entirely on motivation. In laboratory conditions with a target to destroy, a mantis shrimp strikes every 2.5 seconds: 0.4 watts of continuous mechanical power. In the wild, hunting shrimp strike 10 to 20 times per hour, producing perhaps 0.2 watts. For this chapter we assume a motivated shrimp, because an unmotivated shrimp produces nothing except a judgmental stare from sixteen photoreceptors.

Can it Run AI?

With what: mantis shrimp raptorial appendage (spring-loaded)

Output: cavitation-induced mechanical energy

Unit: shrimp-punches (motivated, lab conditions)

Watts: ~0.4W per shrimp (sustained, 1 strike per 2.5 seconds)

Yes. 75 angry shrimp.

Seventy-five mantis shrimp striking continuously in a suitably reinforced aquarium produce 30 watts.

So 1 mantis shrimp = 4,000x Potatoes

and 1 cyclist = 250 Mantis Shrimp

PAINT ME LIKE ONE OF YOUR PEACOCK SHRIMP.

The peacock mantis shrimp (Odontodactylus scyllarus) is known for its vibrant, rainbow-colored exoskeleton, which has a brilliant mix of emerald green, vivid red, orange, and electric blue colors.

Run AI on Pistol Shrimp Swarms

88-92 chromosomes
218 decibels, 5,000 Kelvin

run your AI swarms on shrimp swarms, loudly

When a pistol shrimp snaps its claw shut at 30 meters per second, it creates a cavitation bubble that collapses at 5,000 Kelvin. That is hot enough to ionize water into plasma. The shockwave measures 218 decibels at four centimeters, louder than a gunshot. Colonies snap continuously across entire reefs, masking submarine sonar. Each snap releases 100 nanojoules. We propose scaling this to an inference cluster. The collapsing bubble briefly emits sonoluminescence — visible light from raw mechanical force in seawater. Each shrimp fires 1 to 2 times per second when hunting, producing 150 nanowatts of continuous power. For 30 watts you need 200 million shrimp across 200 hectares of reef, producing inference and deafness in equal measure.

PISTOL SHRIMP ARE LOUDER THAN A GUNSHOT. YOU CAN HEAR THEM FROM SUBMARINES.

Despite being only about 3 to 5 cm (1.2–2.0 inches) long, pistol shrimp make one of the loudest sounds in the ocean by snapping its claws and creating cavitation bubbles that generate, on average, 190 to 218 decibels (dB) underwater.

The bubbles from the pistol shrimp's claws also produce brief, intense heat from 4000 to 5000°C

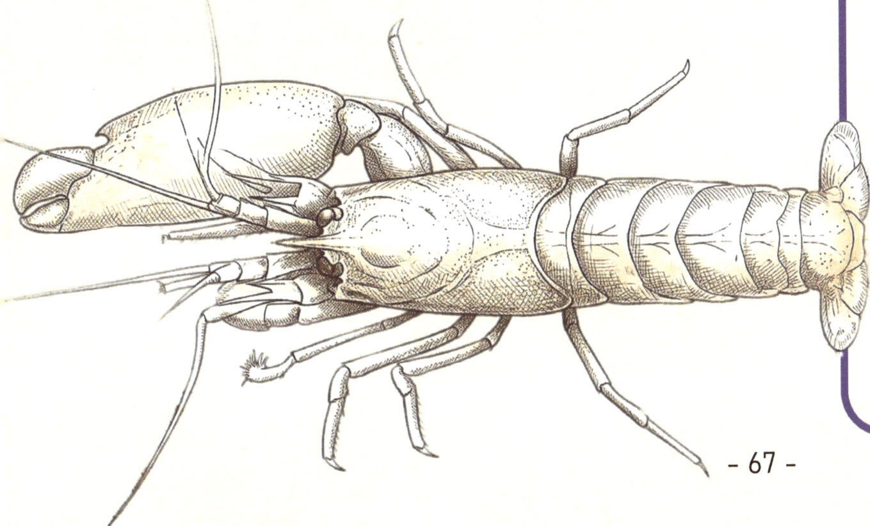

Can it Run AI?

With what: pistol shrimp cavitation bubble collapse

Output: plasma shockwave energy (acoustic + thermal)

Unit: synchronized snaps at 1.5 Hz

Watts: ~0.00000015W per shrimp(150 nanowatts)

Sure, with just 200 million shrimp in a swarm.

One pistol shrimp needs more than 6 years of continuous snapping to make enough energy for a single AI "hello."

So 1 pistol shrimp = 0.0015x Potatoes

and 1 cyclist = 666,666,667 Pistol Shrimp

30-44 chromosomes
usually, some have hundreds

Run AI on Jellyfish

the organism that is a neural net

The jellyfish has no centralized brain. It has a distributed nerve net of roughly 1,000 neurons spread through its bell, with some processing information from 24 eyes on as little as 1 microwatt. When we built artificial neural networks, we accidentally reinvented the jellyfish, scaled it up a billion times, and wondered why the power bill hurt. Aequorea victoria also glows. That green fluorescent protein won Osamu Shimomura the Nobel Prize in 2008. It absorbs blue light at 470 nanometers and re-emits green at 509 nanometers with 79% quantum efficiency. Each flash produces roughly 1 million photons a second. We calculate that at about 4.2×10^{-13} watts of optical power.

Aequorea victoria

Know as the crystal jelly, this bioluminescent hydrozoan jellyfish is found in the Pacific Ocean of North America It contains the photoprotein aequorin and a green fluorescent protein (GFP) involved in bioluminescence.

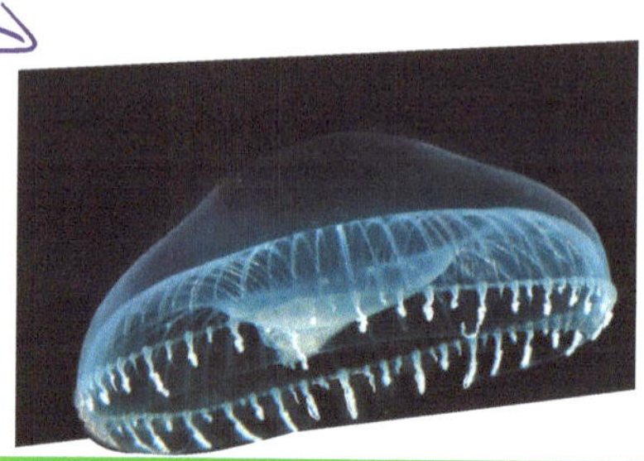

COMPARING NEURAL NETS

Animal	Power	Neurons	Eyes
Jellyfish nerve net	~0.000001W	1,000	24
Human brain	~20W	86 billion	2
GPT-4 (serving)	~3,000W	1.8 trillion param.	0
7B local model	~30W	7 billion param.	0

Note: Approximate, cross-domain comparisons of optical or computational power since no true biological equivalent exists.

Jellyfish process visual information from 24 eyes on 1 microwatt. GPT-4 processes text on three kilowatts and can't see anything. The jellyfish is three billion times more power-efficient per computational decision. It has been doing this for 500 million years.

For AI to Say "Hello" You'll Need about 71 Trillion Jellyfish

To reach 30 watts from jellyfish bioluminescence, you would need 71 trillion jellyfish glowing simultaneously. At one jellyfish per liter, that's 71 trillion liters, or roughly 28 million Olympic swimming pools of continuously luminescent seawater. One jellyfish would need about 2.3 million years to accumulate enough optical energy for a single "Hello," which is longer than modern humans have existed. The jellyfish does not care. It is already running inference on its own neural net, processing visual information from 24 eyes in four color channels, navigating currents, hunting prey, and doing all of it on one microwatt. The jellyfish solved distributed compute 500 million years ago. It's just tough to harness.

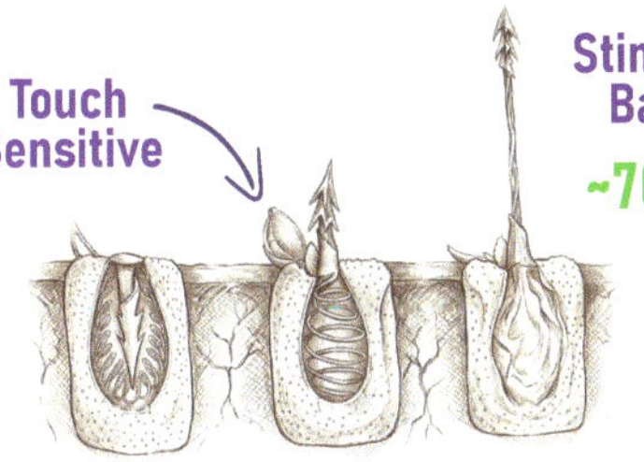

Sting Faster than Bullet from a Rifle

Jellyfish nematocysts are among the fastest mechanical events in biology. Each stinging cell contains a coiled, barbed thread under 150 atmospheres of osmotic pressure. When triggered by contact, the thread fires in 700 nanoseconds, accelerating at 5,400,000 g.

DRIFTING & GLOWING, TOO MANY EYEBALLS, FLOAT MODE ACTIVATED (POSSIBLY BY AI).

When it comes to power, the potato wins over the jellyfish two hundred million to one.

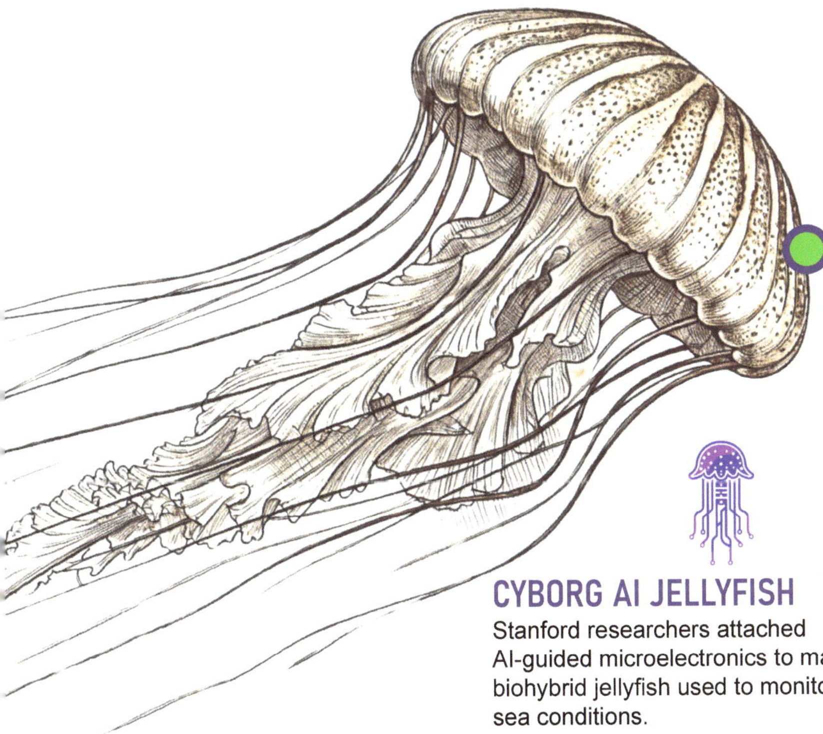

CYBORG AI JELLYFISH

Stanford researchers attached AI-guided microelectronics to may biohybrid jellyfish used to monitor sea conditions.

Can it Run AI?

With: bioluminescent photon
Output: optical emission (509nm)
Unit: jellyfish-hours of glow
Watts: 4.2×10^{-13} W per jellyfish (optical)

Yes, with 71 trillion jellyfish. Glowing. In 28 million pools.

Or one jellyfish, running its own neural net on one microwatt, already doing inference better than your proposal deserves.

So 1 jellyfish makes power of about...
0.0000000042x Potatoes

and 1 cyclist = You would need every jellyfish that has ever lived. It would not be enough.

Just Waves

from the sound of ocean waves

Run AI on Sonoluminescence

energy created by the sound of crashing waves

Sonoluminescence happens in the ocean when acoustic cavitation happens, which is the rapid collapse of air bubbles driven caused by turbulent water motion that creates tiny and incredibly hot flashes of light. Collapse a bubble with sound waves like from turbulent water at the right frequency and it implodes so violently it emits light. The interior reaches 20,000 Kelvin. That is four times hotter than the sun's surface for 100 picoseconds. This is sonoluminescence, tiny bubbles called "acoustic cavitation" producing photons from pressure waves in degassed water. The energy density briefly exceeds stellar cores. The mechanism remains partially mysterious despite decades of study. We propose harvesting these microscopic stellar flares to run inference.

Can it Run AI?

With: sonoluminescence bubble in acoustic standing wave

Output: photon emission from cavitation collapse

Unit: bubble-flashes at 26.5 kHz

Watts: ~2.65×10^{-8} W per stable bubble (26.5 nanowatts)

It'd take 1.13 billion bubbles.

One bubble needs 36 years of continuous flashing to accumulate enough energy for a single AI to say "hello." For real-time inference you need 1.13 billion stable cavitation bubbles making a microscopic star 26,500 times per second.

So 1 bubble is about power of... Potatoes 0.000265x

and to get the power of 1 cyclist...
3,770,000,000
Bubbles are needed

A single bubble driven by a 26.5 kilohertz standing wave collapses to one-hundredth its diameter in under a nanosecond, releasing one picojoule per flash. The bubble oscillates stably at 26,500 flashes per second, producing 26.5 nanowatts of continuous power. That's roughly one four-thousandth of a potato. A tiny star, 26,500 times per second, weaker than a tuber rotting in dirt. For 30 watts you need 1.13 billion stable bubbles, each in its own tuned acoustic chamber. The apparatus would be magnificent. The power bill for the transducers would dwarf the output by several orders of magnitude.

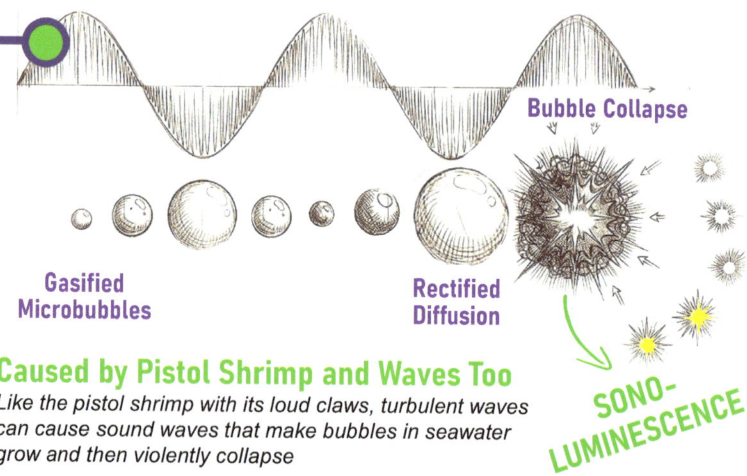

Caused by Pistol Shrimp and Waves Too

Like the pistol shrimp with its loud claws, turbulent waves can cause sound waves that make bubbles in seawater grow and then violently collapse

4x HOTTER THAN THE SUN. FOR JUST A MOMENT.

Run AI on Dinoflagellates

bioluminescent ocean waves, visible from space

VARIES
500 million years of diversity

Dinoflagellates are a diverse group of single-celled eukaryotic organisms, primarily marine phytoplankton (algae) that use their flagella to swim and produce light through a luciferin-luciferase reaction housed in specialized organelles called scintillons. When mechanically disturbed by waves, a passing fish, or your hand, their calcium channels open to produce a 100-millisecond flash of blue light at ~470 nanometers. The quantum efficiency is roughly 20 percent, meaning one in five chemical reactions produces a photon rather than just heat. This is respectable for biology. The problem is power. A single flash releases one ten-billionth of a joule.

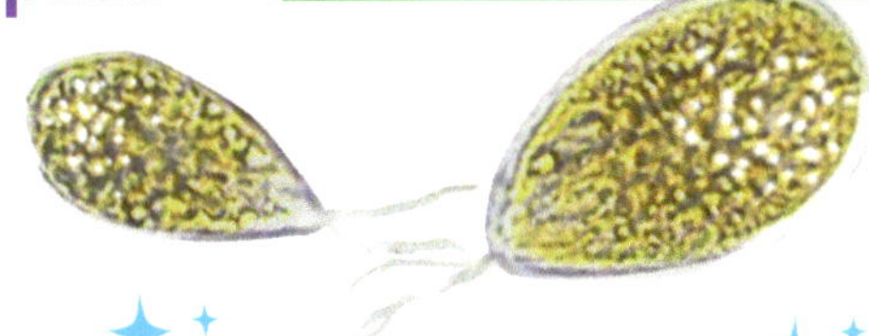

Ostreopsis: Toxic beauty

Ostreopsis is a genus of marine benthic dinoflagellates (microalgae) that are found across the world in tropical and temperate coastal waters. They form harmful blooms, especially during summer, producing toxins (palytoxins) and known for their glow.

MAKING SEAS GLOW FOR 650 MILLION YEARS.

There are ~18 bioluminescent genera, with (depending on who you ask) 60 to 100 glowing species. That's a small fraction of the 2,000 to 4,500 living species of dinoflagellates.

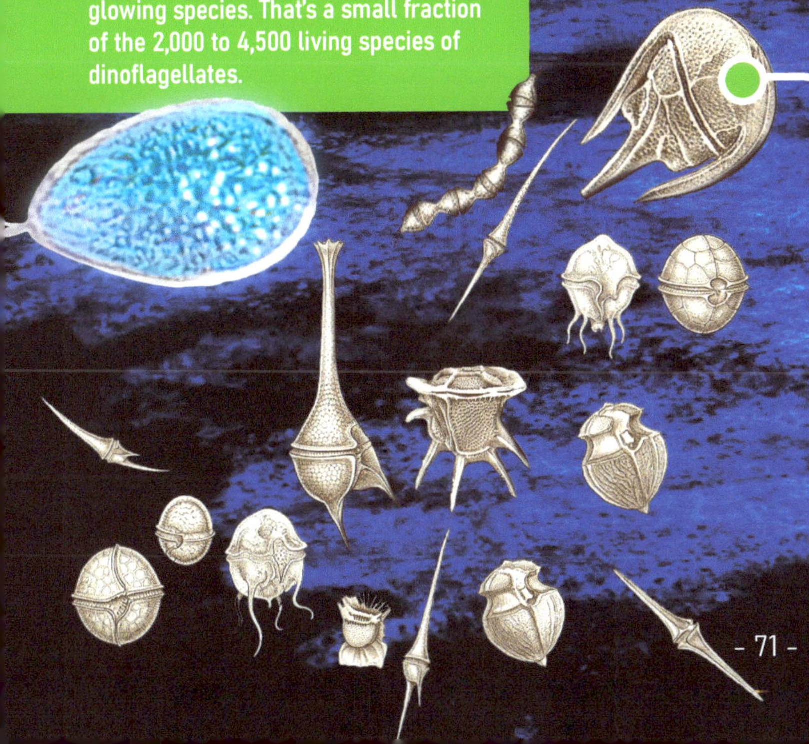

Can it Run AI?

With: dinoflagellate bioluminescent flash (luciferin-luciferase)

Output: blue photons at 470 nm from mechanically stimulated cells

Unit: agitated cells at 0.1 Hz

Watts: ~10^{-11} W per cell (0.00000000001 W)

Yes, on just 3 trillion dinos.

One dinoflagellate needs ~95,000 years of continuous flashing to power a single "hello," good thing oceans contain trillions.

So 1 dinoflagellate is power of about...

0.0000001x Potatoes

and the power of 1 cyclist takes...

10,000,000,000,000 Dinoflagellates

6-40 chromosomes
depends which of the quintillion

What Else Makes 500 GW?

Source	Output
All US electricity generation	~450 GW
500 nuclear power plants	~500 GW
5 million small wind turbines	~500 GW
5 quadrillion potatoes	~500 GW
Every copepod in the ocean	~500 GW

One of these is barely visible to the naked eye.

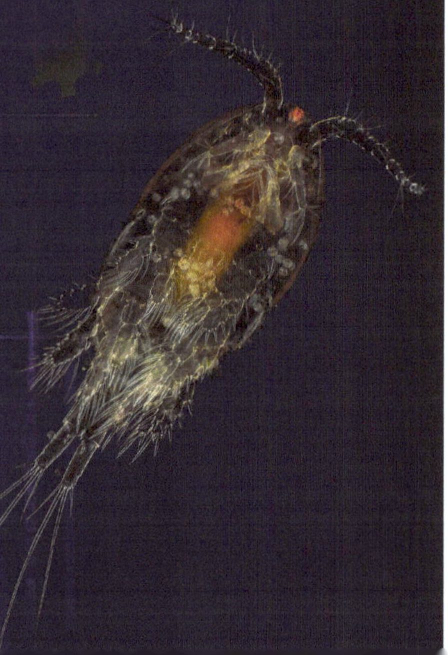

Run AI on Copepods

individually nothing, collectively everything

A copepod is a crustacean smaller than a grain of rice. Most are between half a millimeter and two millimeters long, weigh about five micrograms, and produce approximately half a microwatt of metabolic power. That's 0.5% of a potato. Half of one percent. A single copepod is the least powerful organism in this book by a comfortable margin. It has no brain, no centralized nervous system, and no opinion about inference. It eats algae, reproduces, and drifts. One copepod, running continuously, would need about 1.9 years to accumulate enough power for one AI to say 'hello.' But it'd look really cute in the jar while doing it. And if you have a saltwater aquarium, you probably already have thousands.

The Scale Table

Scale	Count	Power	Potatoes	AI Hello's
1 copepod	1	0.5 µW	0.005x	not in your lifetime
A jar	1,000	0.5 mW	5x	not in your lifetime
An aquarium	10,000	5 mW	50x	still no
A pond	1,000,000	0.5W	5,000x	one in 60 seconds
1 km^2 of ocean	1,000,000,000	500W	5,000,000x	17
All the oceans	10^18	500 GW	5×10^{15}	17 billion

The jump from "a pond" to "all the oceans" is nine orders of magnitude. Somewhere in there, the copepods stopped being a joke and started being infrastructure.

Copepods from Ernst Haeckel's Kunstformen der Natur (1899)

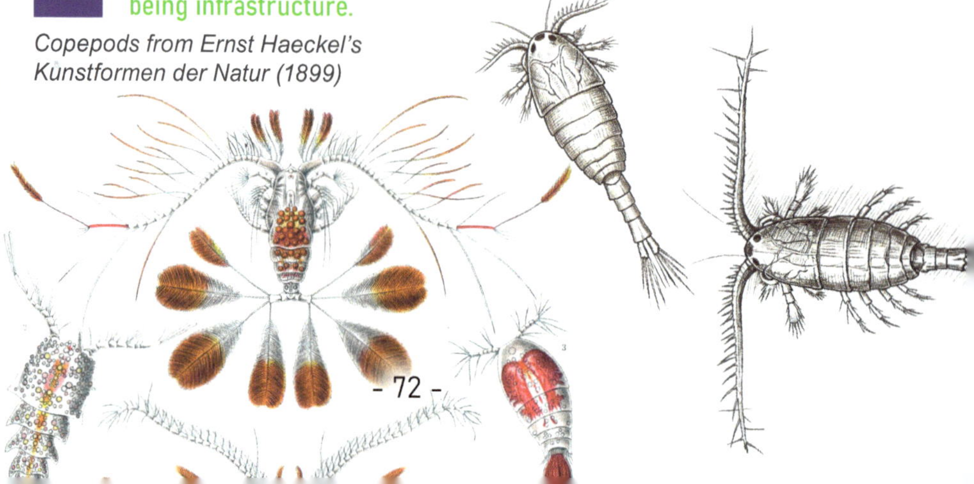

How Many Copepods are in the Ocean?

There are an estimated one quintillion copepods in the ocean. That's 10^{18}, or roughly ten million copepods for every star in the Milky Way. Their combined metabolic output is approximately 500 gigawatts. That's more than the entire United States electricity consumption. Five hundred nuclear power plants. Seventeen billion simultaneous AI conversations. The ocean is already generating more biological power than every data center, solar farm, and wind turbine on the continent combined, and it's doing it with animals you can't see without a magnifying glass. Every chapter in this book asks "can it run AI?" Copepods are an organism where the answer is, individually, absolutely not. Collectively, they could run all of it.

Copepods were first observed in the late 17th century. Antonie van Leeuwenhoek revolutionized the microscope and wrote about them in 1699.

AN OCEAN-SIZED DATA CENTER. JUST HARD TO GET COPEPODS TO STAY IN RACKS.

More Copepods Than Stars

There are approximately 100 billion stars in the Milky Way. There are approximately 10 quintillion copepods in the ocean. That's 10 million copepods per star. If each copepod were a pixel, you could render the entire observable universe at high resolution and still have copepods left over. They are the most abundant multicellular animals on Earth. They have been here for 200 million years.

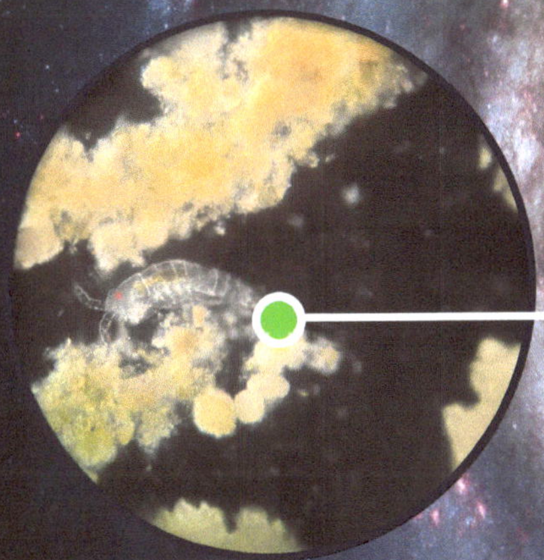

Saltwater copepod at 200x magnification
(my home marine aquarium)

Can it Run AI?

With what: `metabolic heat + kinetic motion`

Output: `metabolic power`

Unit: `copepods`

Watts: `0.5 µW each (that's 0.0000005W per copepod)`

Only if we connect all of them.

One copepod produces 0.5% of a potato. One quintillion copepods produce 500 GW. The math is linear. The implication is not.

So 1 copepod produces about the power of...

`0.005x`
Potatoes
(potato wins by 200x)

and 1 cyclist = `200,000,000x`
Copepods
(but there are 10^{18} of them, so this is a rounding error)

Section 7

Creepy Crawlies

small things with a big impact, & AI is already reshaping how we see them.

The creepy crawlies of the world pollinate our flowers, decompose our trash, and keep the leaf litter alive. And some of them evolved power systems stranger than anything in your house. Bombardier beetles spray boiling chemicals. Oriental hornets harvest sunlight through cuticles. Bees heat their hives like defensive reactors. Spiders plug directly into the Earth's electromagnetic field. Mushrooms and slime mold get uncomfortably close to computation.

We have lived with these creatures for thousands of years and still do not fully understand them. AI is helping us look closer by tracking invasive species, spotting locust swarms before they devour crops, and revealing patterns in biological systems too small or too numerous for us to follow alone. In agrirobotics, AI is even enabling us build robots base don insects. They could run our models, but our models may help us finally understand theirs.

Run AI on Bombardier Beetles

the power of exploding beetles, that's a good one

35 chromosomes
and it really explodes

The bombardier beetle built a chemical cannon. Hydrogen peroxide plus hydroquinone, detonated in a reinforced abdominal chamber, fired through a 270-degree rotating turret at 100°C and 500 pulses per second. Peak output: 78.5 watts, a laptop charger with targeting capability. The problem is ammunition. This beetle has approximately 40 shots in reserve. At three encounters per day, sustained spray output collapses to almost nothing. The beetle's metabolic heat, about 5 milliwatts. It is an outstanding weapon. Terrible stamina though.

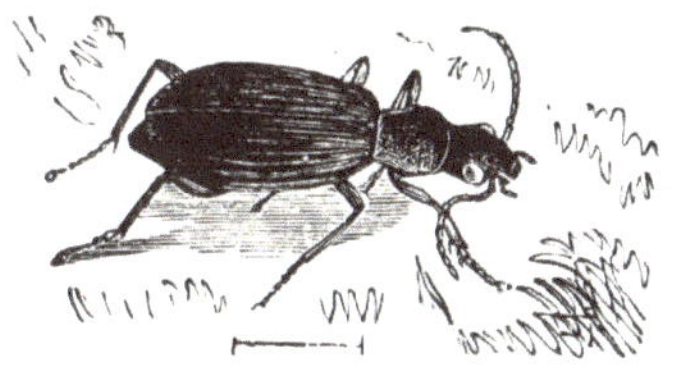

BOMBARDIER BEETLE.—*Brachinus crepitans.* (Magnified.)

An llustration from B.A. D'Abreu. Beetles of the Himalays (1885) book on beetle species

NATURE ENGINEERED ITS OWN EXPLOSIVES AGENT.

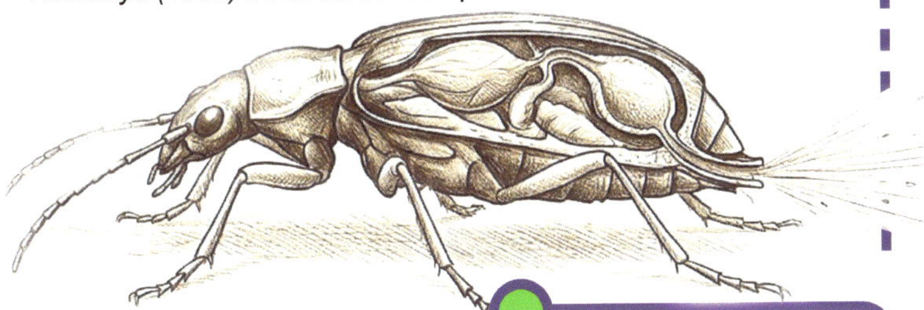

Can it Run AI?

With: exothermic chemical spray

Output: 100°C quinone pulses

Unit: ~0.005W sustained

Watts: 78.5W
(while ammunition lasts)

6,000 beetles. Limited ammo.

The spray is spectacular but episodic. The metabolic heat is boring but continuous. The boring number is the real one.

So 1 beetle has the power of about...

50x
Potatoes

and 1 cyclist's power takes about...

20,000
Beetles

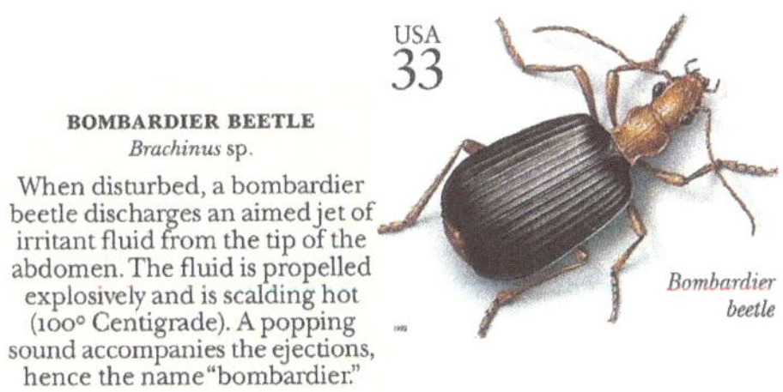

In 1999 the U.S. Post Office issued a stamp with the image of a Bombardier beetle. What's so special about this insect? Well firstly this beetle predates humanity, and it is the only life form other than humans that creates explosions.

32 chromosomes
females (males have 16)

Run AI on Oriental Hornets

the only animal that is a solar panel

Oriental hornet discovered photovoltaics 25 million years before Bell Labs. Its yellow cuticle contains xanthopterin, a pigment that converts sunlight to electrical potential. Its brown bands contain melanin arranged in a hexagonal lattice with 15-nanometer spacing, small enough for quantum effects, functioning as a biological semiconductor. Total output: approximately 1 microwatt per hornet in full sun. You need 30 million of them, all sunbathing simultaneously, to reach inference threshold. Evolution had better aesthetics. Silicon has better power density.

Can it Run AI?

With what: `xanthopterin/melanin photovoltaic cuticle`

Output: `photoelectric current`

Unit: `hornets (in sunlight)`

Watts: `~0.000001W (1 µW)`

Sure, it only takes about 30,000,000 hornets sunbathing.

It's unique in that it generates electricity from light through solid-state physics, not chemistry.

So 1 hornet = 0.01x Potatoes

and 1 cyclist = 100 million Hornets

AI NOW TRACKS INVASIVE HORNETS

VespAI's AI-powered system developed by the University of Exeter can detect & track invasive Asian hornets with >99% accuracy.

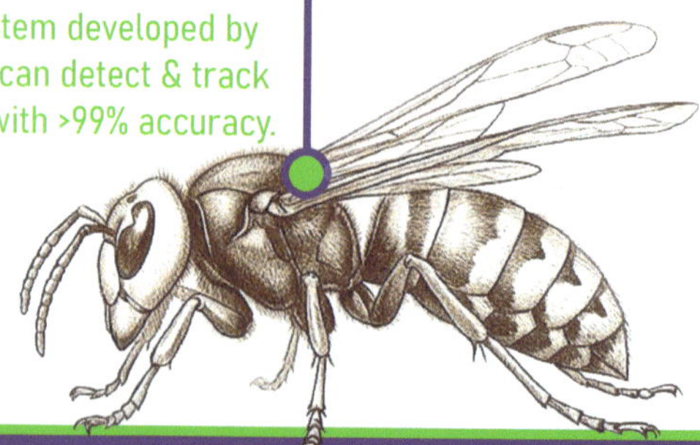

HI, I'M MRS. HORNET. I'M BASICALLY A SOLAR PANEL.

Oriental hornets are eusocial insects with strict division of labor between queens, workers, and males, who carry half the chromosomes. Females do everything: build the nest, run the colony. Males only exist briefly to reproduce. The female's unique photovoltaic cuticle acts like a solar panel. The solar panel is female.

Run AI on Bee Swarms

it's not just one you need to watch out for

A single honeybee generates heat by vibrating its flight muscles with its wings folded. At peak output that generates 0.11 watts, briefly, before it burns out. But bees don't work alone. A winter hive of 50,000 bees shares the load, each contributing a tiny amount of heat to keep the whole cluster at exactly 35°C (95°F). Together they produce 150 watts, enough for AI to say "hello" 5 times.

One bee's wing vibrations also generate a small static charge of about 0.1 milliwatts, exactly one potato of power. When threatened, bees can pile on an intruder and vibrate together until the temp reaches 47°C (117°F), hot enough to kill a hornet.

32 chromosomes
females (males just 16 again)

HONEYBEE SWARM

In transit between hive locations, ~10,000 to 50,000 bees make heat.

AI TRACKS BEES

AI tech like Beewise and Ubees can use AI vision to count, track, & monitor bees.

AI IS MOVING FAST. IT'LL BEE OK.

POLLINATION BY AI-DRIVEN ROBOT BEES

RoboBee research focuses on creating AI-driven robotic bees that can pollinate plants, primarily developed by teams at Harvard University, MIT, and industrial researchers affiliated with Japan's National Institute of Advanced Industrial Science & Technology.

It's Getting Hot in Here (too hot for hornets)

Bees generate heat when swarming, which can help to protect their hives, getting as hot as 47°C (117°F) hot enough to kill other insect intruders, like attacking hornets.

Can it Run AI?

With what: thoracic muscle thermogenesis
Output: metabolic heat
Unit: bees (swarming)
Watts: ~0.003W per bee (cluster) 0.11W peak

Yes, 10,000 bees in a swarm.

Or 273 bees at peak active heating, for the few minutes before they overheat and die. The sustained number is the real one. One hive produces 5 conversations.

So 1 bee has the power of about...

30x Potatoes

and 1 cyclist's power takes about...

33,333 Bees

18 chromosomes
some species have 22

Run AI on Fireflies

twinkle twinkle little... wait is that a robot?

Fireflies make light at 88% efficiency through a chemical reaction in their abdomen — better than any LED we've built. Each flash lasts 200 milliseconds and produces 2.5 microwatts. You need 12 million of them, all flashing simultaneously, to reach inference threshold. The more impressive number is the synchronization. Thousands of fireflies in the same canopy will start flashing independently, watch their neighbors, and within minutes lock into a single coordinated pulse. No leader, no clock, no WiFi. Just local observation, repeated until the whole forest agrees.

Can it Run AI?

With: bioluminescence
Output: green light (560 nm)
Unit: fireflies (during display)
Watts: ~0.0000025W (2.5 µW per firefly)

Yes, 12,000,000 fireflies.

The synchronization is more impressive than the power output.

So 1 firefly = 0.025x Potatoes

and 1 cyclist = 40M Fireflies

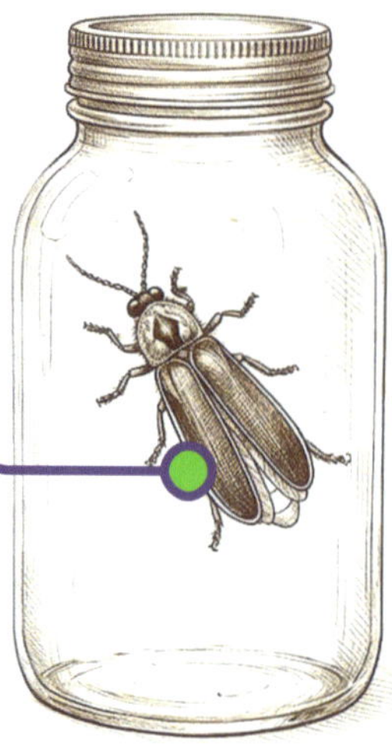

EFFICIENCY COMPARISON

Light Source	Efficiency
Incandescent bulb	~5%
Fluorescent tube	~22%
White LED	~50%
Firefly	~88%

AI FIREFLY GLOWBOTS

The firefly is one of the most efficient light sources on Earth. MIT researchers made tiny AI/ML powered artificial muscles that flap robotic insect wing and have electroluminescent particles emitting colored light during flight, similar to fireflies.

Fireflies in south Georgia, U.S., lighting up the night at the Butler Island Plantation (8-second exposure captures 5 to 6 firefly flashes)

MOST EFFICIENT LIGHT SOURCE ON EARTH.

Run AI on Locusts

a swarm of low-cost power solutions

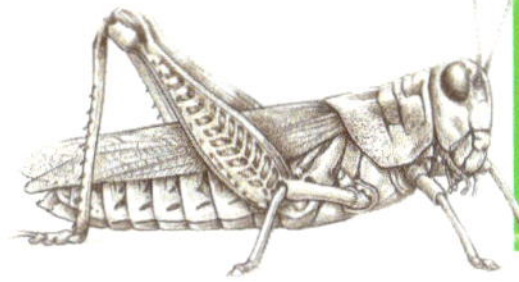

22 chromosomes
biblically confirmed

A single desert locust in flight burns 0.25 watts, running its flight muscles at 30 beats per second while eating its own body weight every day. You need 120 locusts to reach inference threshold. That's a shoe box of flying locusts. The problem is locusts don't fly in shoe boxes.

Locust Swarms and Powering AI

A small swarm is 40 million locusts producing 10 megawatts. The 2020 Kenya outbreak reached an estimated 200 billion across multiple swarms: 50 gigawatts of collective metabolic power, enough for 1.67 billion simultaneous conversations. Each locust just follows simple rules, avoid being eaten, eat everything else, keep up. Nobody is in charge. The continent gets devoured anyway.

AI LOCUST DETECTION

Agribots (or agriculture robots) researchers led by Najran University in Saudi Arabia used AI and ML to detect pest locusts in fields with up to 99.51% accuracy, aiming to help farmers save crops.

Bizarre locust engraving (1594) depicting real and imagined "demon" locusts in a field of insects

Can it Run AI?

With: metabolic flight power
Output: sustained aerobic watts
Unit: locusts (in flight)
Watts: ~0.25W per locust

So 1 locust = 2,500x Potatoes

and 1 cyclist = 400 Locusts

Yes, we need 120 locusts in a box.

Or 40 million in a small swarm producing 10 megawatts. The individual number is achievable. The collective number is biblical. Neither generates electricity.

SEEN THROUGHOUT HISTORY

Locusts: Apparently the "Demon" We All Had in Common Across the Ancient World

Locusts appear associated with judgment or gods (and sometimes demons or other deities) across the Bible, Quran & Hadith, Sumerian and Akkadian texts, Egyptian texts and in Pharaohs' tombs, and even in Native American oral traditions, with remarkable consistency.

Drawing shows locust detail from a hunt mural in the grave chamber of Horemhab, last pharaoh of Egypt's 18th Dynasty, c.1422–1411 BC. Over 3,400 years old and still recognizable.

7-128 chromosomes
nobody can exactly agree

How the Field Works

Earth's surface carries a net negative charge. The ionosphere carries a net positive charge. Between them: a fair-weather electric field of 100-150 volts per meter pointing downward. Lightning recharges this system continuously, striking the ground roughly 100 times per second worldwide. The spider's negatively charged silk is repelled upward by the negatively charged ground. The entire atmosphere is a capacitor and the spider is exploiting the leakage current. This has been working for 400 million years. It requires no maintenance.

Run AI on Spider Ballooning

surfing the planet's electromagnetic field on silk

Spiders fly without wings. They climb, raise their abdomens, release silk into the air, and lift off. In 2018, Morley and Robert demonstrated that spiders launch in zero-wind conditions by exploiting Earth's atmospheric electric field, about 100 volts per meter near the ground that pushes up on negatively charged silk. The spider becomes a biological capacitor coupled to a planetary gradient. Each strand charges to roughly 1 nanocoulomb, producing about 40 nanowatts of sustained coupling. You would need 750 million ballooning spiders to get your AI to say "hello." That's approximately 52,000 per square kilometers, spread across the entire state of Connecticut.

Distance Records

Spider Ballooning Record	Distance
Typical dispersal	1-10 km
Documented maximum	1,600 km (collected on ships mid-ocean)
Altitude	Up to 4-5 km (collected on aircraft)
Charles Darwin (1832)	Noted "aeronaut spiders" landing on HMS Beagle 100+ km from shore

Darwin watched them land on the Beagle and wrote about it in The Voyage of the Beagle. He assumed wind. It took us 186 years to discover it was electricity.

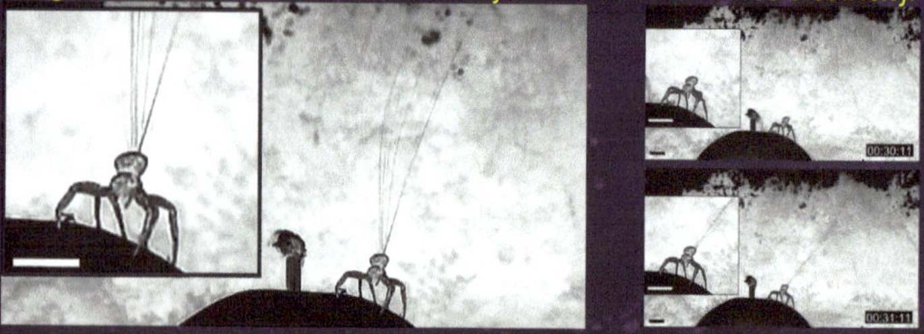

AI IS HELPING RESEARCHERS UNDERSTAND SPIDERS

AI/ML was used by MIT researchers to classify spider web vibrations into musical "orchestras" that predict behavior, and at Johns Hopkins researchers used neural networks to track web building behaviors.

How it Works: Electric Parachutes

The energetics are almost comically small. Each silk strand charges to roughly 1 nanocoulomb at a potential difference of 200 volts across its 2-meter length. Energy per launch is 200 nanojoules. If the spider could cycle this every 5 seconds, sustained coupling to Earth's field produces roughly 40 nanowatts. You would need 750 million ballooning spiders to reach 30 watts. At that density, you'd need roughly 52,000 spiders per square kilometer, or about one for every parking space across the entire state of Connecticut. But spiders already balloon in enormous numbers: millions launch daily across temperate regions, drifting hundreds of kilometers on invisible currents. They were here before the dinosaurs. They've been riding the planet's electrical field for 400 million years. The power is negligible. The mechanism is extraordinary. Every other organism in this book generates its own energy. The spider plugs into the Earth.

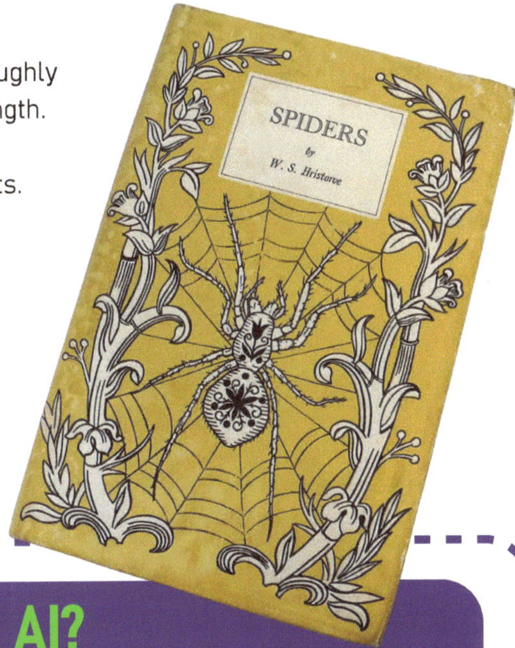

Can it Run AI?

With what: electrostatic coupling to atmospheric field
Output: charged silk in 100 V/m gradient
Unit: ballooning events (5-second cycles)
Watts: ~4×10^{-8} W per spider (40 nW)

750,000,000 spiders. Ballooning. On silk threads.

The power is irrelevant. The physics is the story. Every other power source in this book is self-contained. The spider is the only organism that uses the planet itself as its power supply.

So 1 ballooning spider = 0.0004x **Potatoes** (the potato wins by 2,500 to 1)

and 1 cyclist = 2,500,000,000 **Spiders** (two and a half billion)

SHE SWALLOWED A SPIDER TO CATCH THE FLY, AND I DON'T KNOW WHY, MAYBE ASK AI.

The termites ate this page.

A colony of 60,000 pest termites can consume about one foot of a 2x4 beam in roughly five months.

We suspect they consumed this page and enjoyed it.

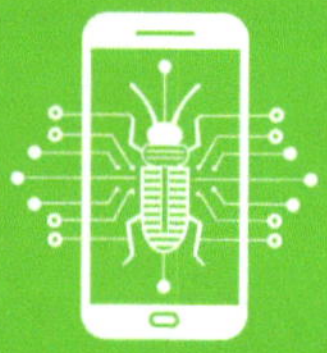

TERMITE ID WITH AI/ML

But thanks to AI/ML researchers in Taiwan at National Chung Hsing University (NCHU) and collaborators at Academia Sinica we can identify which termites did it using deep convolutional neural networks (DCNNs) to identify termite species right from a smartphone.

Run AI on Termite Mounds

28-56 chromosomes
from queens to workers

mounds of power when all of them get together

A macrotermes mound contains roughly one million termites, each producing 20 microwatts of metabolic heat. Collectively: 20 watts. That is two-thirds of the power needed to get to your AI to say "hello" from insects that weigh 3 milligrams each. Two mounds would get you to AI "hello." But the mound does something stranger than generating power. It regulates its own internal temperature to within 1°C despite external swings of 40°C. It does this through passive ventilation, a network of chimney channels, intake tunnels, and thin-walled gas exchange structures that adjust airflow based on thermal gradients.

No thermostat. No sensors. No central control.

The architecture itself solves differential heat equations through physics. The termites built it with no blueprint, no foreman, and no knowledge of fluid dynamics. Each termite follows local chemical rules. The structure emerges. The computation is the building. This is the only chapter where the power source and the computation are the same system: one million organisms generating 20 watts of heat inside an architecture that computes what to do with it. It is not inference. But it is the closest anything alive has come to running its own data center.

TERMITES OFTEN BEAT AI AT MATH, SINCE THEY KNOW LOGS.

Can it Run AI?

With: heat (Macrotermes)

Output: thermal energy

Unit: mounds (~1 million termites)

Watts: ~0.00002W per termite (20 μW)

Yep, 1.5 Mounds.

One mound gets you to two-thirds of inference threshold. The mound also computes its own climate. It just can't compute yours.

So 1 mound is the energy of...

200,000x Potatoes

and 1 cyclist's power is about...

5 Termite mounds

1 chromosome
and its circular

Run AI on Microbial Fuel Cells (MFCs)

one man's waste is one microbe's energy

Geobacter sulfurreducens eats electrons from its environment and excretes them through protein nanowires: actual conductive filaments micrometers long that function as biological USB cables. Feed it organic waste on one electrode, collect electrons on the other, and the bacteria generate current indefinitely, because they're not trying to power your inference. They're trying to stay alive. The current is metabolic exhaust. A well-designed microbial fuel cell produces 0.3 watts per square meter of anode surface. Under optimized laboratory conditions with engineered strains: 1.5 watts per square meter. For 30 watts at typical performance, you need 100 square meters of colonized anode: a 10-meter by 10-meter room of electrodes, fed wastewater, smelling exactly like you'd expect. It runs indefinitely. It costs almost nothing. It smells terrible. This is the most honest power source in the book.

Can it Run AI?

With what: `bacterial electrons`

Output: `DC nanowire excretion`

Unit: `square meters of colonized anode`

Watts: `~0.3 W/m² (typical)`

Yes, 100m² biofilm anode.

The bacteria don't know and don't care. They are eavesdropped metabolism. This may be the most achievable biological power source in the book.

So one biofilm anode...

3,000x **Potatoes**

and 1 cyclist is about a tennis court of bacteria...

333m² **Biofilm anode**

Digestion is Energy

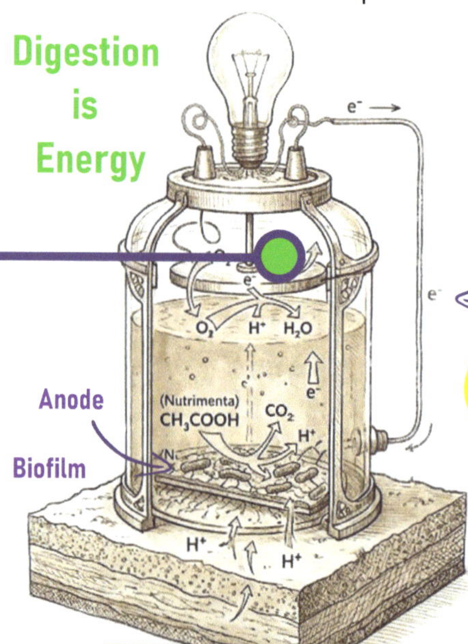

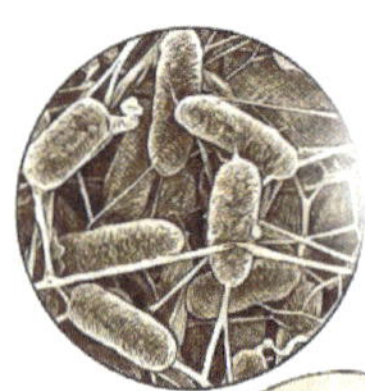

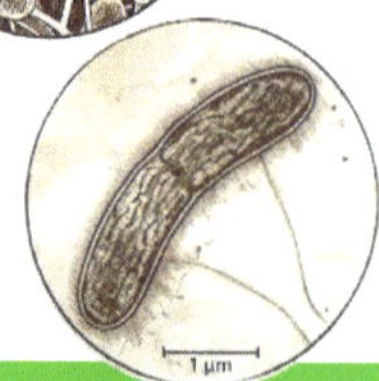

AI FOR FEEDING MFCs

An AI system named EvoBot feeds MFC bacteria, accelerating maturation of energy-producing cultures from 30 days to 6 days.

RUNS ON SEWAGE. SMELLS LIKE SEWAGE.

A microbial fuel cell (MFC) is a bioelectrochemical system that generates electricity using natural metabolic activity of microorganisms to break down biodegradable materials (like sewage) at an anode, releasing electrons to a cathode through an external circuit & making clean power.

Run AI on Mushrooms

he's got energy, they always say he's a fun guy

Oyster mushrooms generate electricity by decomposing wood chips. Researchers at the University of the West of England discovered that Pleurotus ostreatus, while quietly digesting cellulose, produces electrons through metabolic respiration that flow to external electrodes. One milliwatt per fruiting colony, sustained for weeks, as long as substrate remains. Unlike potatoes, which merely conduct electricity, mushrooms generate it through digestion. Unlike hamsters, they require no motivation. They exist, they decompose, they leak electrons. You would need 30,000 colonies to reach inference threshold. The mycelial network spanning kilometers underground communicates via electrical pulses, exhibits memristive properties, and can solve pathfinding problems. The mushroom is simultaneously a power source, a communication network, and a primitive computer. It does all three badly. It has been doing all three for 400 million years without your help.

It Varies

many different genomes

ACTUALLY MUSHROOMS CAN RUN AI, WHO KNEW

A team of researchers from Unconventional Computing Lab, UWE Bristol founded by Andrew Adamatzky published that fungi are memristors (resistor with memory), a universal logical element in bioelectronics that could feasibly one day run AI.

The underground bacteria and mushrooms all generate power AND compute. They just compute different things than you want.

Mushroom mycelium is the vegetative, root-like network of fungi, composed of thread-like filaments called hyphae that grow underground or within substrates like wood.

Can it Run AI?

With what: metabolic electrons
Output: low current electron flow
Unit: fruiting colonies
Watts: ~0.001W per colony

Yes, with 30,000 mushroom colonies.

The mushroom produces ten potatoes of power. The mushroom is also, itself, a computer. It just can't be redirected to your problems.

So 1 colony = 10x Potatoes

and 1 cyclist = 100,000 Colonies

100s of chromosomes
and wow do they vary wildly

What tests has slime mold solved?

Test	Time	Ref.
Tokyo Rail Map	26 hrs	Tero et al.
Maze Path	4-8 hrs	Nakagaki et al.
Steiner Tree	Hours	Multiple srcs.
US Interstate	24 hr	Adamatzky et al.
Foraging	n/a	Reid et al.

**No brain.
No neurons.
No central processing.**

It has no brain. It has no neurons. It has no central processing. It remembers where it has been by leaving chemical trails. It computes by existing.

Run AI on Slime Molds

finally an organism that actually computes

Every chapter in this book asks "can it run AI?" The answer has been no, creatively, for over a hundred chapters. The slime mold is different. Physarum polycephalum is a single-celled organism that grows as a network of tubes, extending pseudopods toward food sources and retracting from dead ends.

As it explores, the network self-optimizes. Tubes carrying more flow get wider, tubes carrying less get pruned. The result is a minimal spanning network connecting all food sources. In 2010, researchers placed oat flakes on a map of the Greater Tokyo Area at positions corresponding to major stations. The slime mold grew a network that approximated the actual Tokyo rail system. It solved a transportation optimization problem that took human engineers decades using analog chemical computation where the organism's body becomes the answer. But not this book's answer. Slime molds may know shortest paths, minimal networks, efficient routing, but it cannot multiply matrices. It will never run a forward pass. The only organism that actually computes turns out to compute the wrong thing. The spec card still measures watts, because that's all we have.

USING CHEMICAL TRAILS IT CAN INTELLIGENTLY SOLVE MAZES & MORE.

Physarum polycephalum: The "Many Headed" Slime Mold

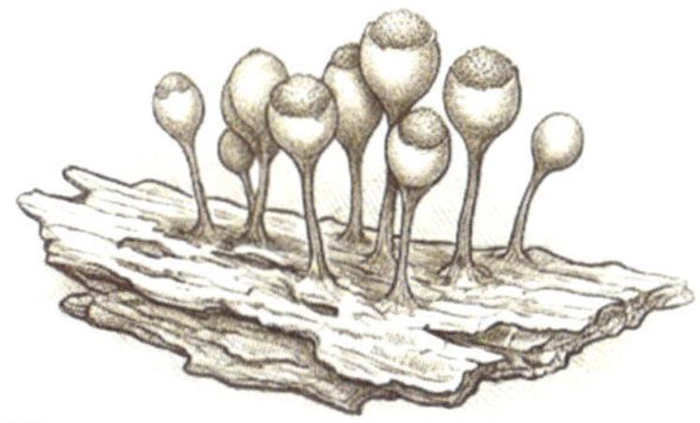

Physarum polycephalum is a bright yellow, single-celled, multinucleate protist (acellular slime mold) that lives in moist, dark environments, and feeds on microorganisms.

A 100 gram laboratory culture of Physarum, roughly the size of a dinner plate, produces 0.1 watts of metabolic heat during active growth and network formation. For 30 watts of inference power, you need 300 cultures totaling 30 kilograms of slime mold. But power is the wrong metric for this organism. The slime mold is not generating energy for computation. It is the computation. The network formation itself is the algorithm. The tubes become the shortest path. The organism doesn't solve the problem and then output an answer. The organism's body becomes the answer to intelligence.

Not Just Power - It computes!

While its calculations may be chemical rather than neuronal, it shows foraging ability consistently, as well as ability to solve complex mazes.

Can it Run AI?

With what:	`cytoplasmic streaming`
Output:	`metabolic heat + topological computation`
Unit:	`grams of active culture`
Compute:	`~0.001 W/g (metabolic)`

Yes. It Actually computes.

For power, 300 cultures produce 30 watts. For computation, one culture solves complex pathfinding problems in hours. These are different questions and the second one is more important.

So 1 dinner plate (~100g) of slime mold is producing power of...

1,000x Potatoes

and 1 cyclist's power takes about...

100kg Slime mold

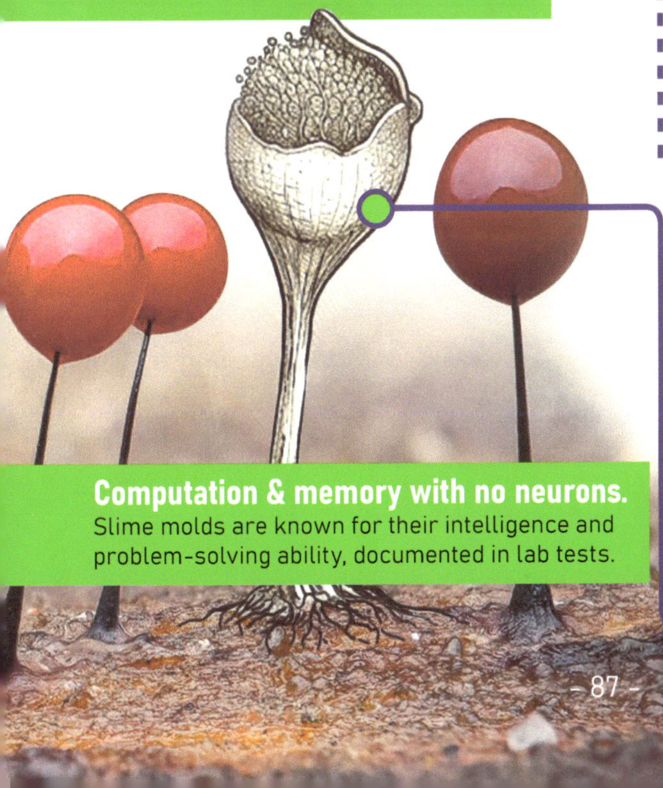

Computation & memory with no neurons.
Slime molds are known for their intelligence and problem-solving ability, documented in lab tests.

Part 3

THE BODY AS COMPUTE

Section 8

Human Power

your body is a generator. a terrible, inefficient, beautiful generator.

Your body is a power plant that runs on sandwiches. You are generating watts right now, reading this sentence, and there is nothing you can do about it. You produce 82 watts lying in bed. You produce 30,000 watts as cremation exhaust, once, at the end, whether you planned to or not. Everything between those two numbers is this section.

Every chapter before this looked outward. This section points the harvest inward. We are looking for power from you, especially when you are not expecting to give it. You generate it snoring loudly enough to shake a piezoelectric membrane. You generate it chewing gum long after the flavor is gone. You generate it radiating heat into your pillow. You generate it making more humans. And you generate one final, massive burst by leaving.

This section would make 'The Matrix' proud. Because in strictly thermodynamic terms, we can run AI on you.

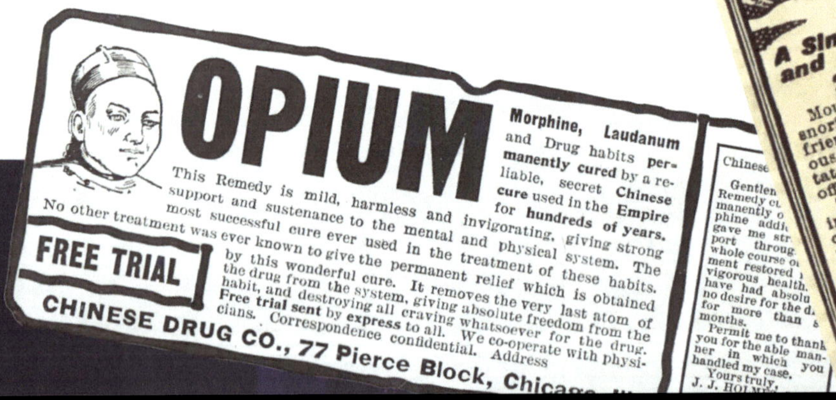

Run AI on Snoring

nocturnal noise power from your pillow

300+ calories per night
making energy while you sleep

A severe snorer generates approximately 55 to 80 decibels at one meter (and sometimes up to 110 decibels), ranging from the volume of a garbage disposal to low flying jet plane. Snoring is mechanical energy, with compression waves propagating through bedroom air, vibrating walls, rattling nightstands, and ending marriages. We propose harvesting it with a bedside piezoelectric membrane.

Snores drop power, bedmates don't care

Sound is pressure oscillation. At 80 decibels, the acoustic intensity reaching a one-square-meter collector is approximately 100 microwatts. A piezoelectric membrane tuned to the fundamental frequency of soft palate vibration, roughly 120 Hz, converts about 5% of this to electrical current (about 5 microwatts). Sleep studies show approximately 60% duty cycle for habitual snorers, with quiet breathing phases interrupting the harvest (average 3 microwatts).

SOMETIMES AS LOUD AS A JET PLANE.

12% of people snored louder than 55 decibels, which is the max legal outdoor nighttime noise in the state of Minnesota, says Karger Publishers.

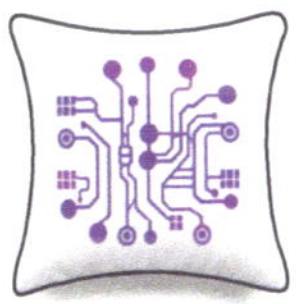

AI SMART SLEEP

AI-enabled smart pillows & mattresses from 10 Minds, DeRUCCI, Nitetronic, and more promise snoring relief and better sleep.

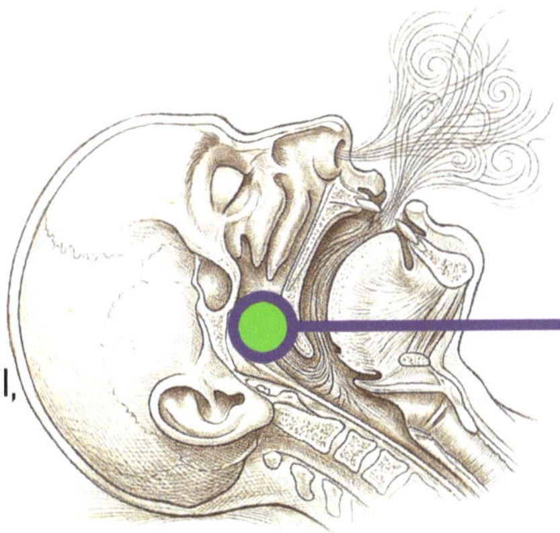

Can it Run AI?

With what: Severe habitual snorer + bedside piezo membrane

Output: Acoustic pressure wave harvesting at 120 Hz

Unit: snore-watts

Watts: 0.000003W (3 microwatts, 60% duty cycle)

Yes. 10,000,000 snorers.

An average apartment building has 40 habitual snorers producing 0.00012 watts collectively. This is not enough to run AI practically. It is enough to run a marriage counselor's waiting room clock.

So 1 snorer makes the power of... 0.03x Potatoes

and 1 cyclist is about... 33,333,333 Snorers

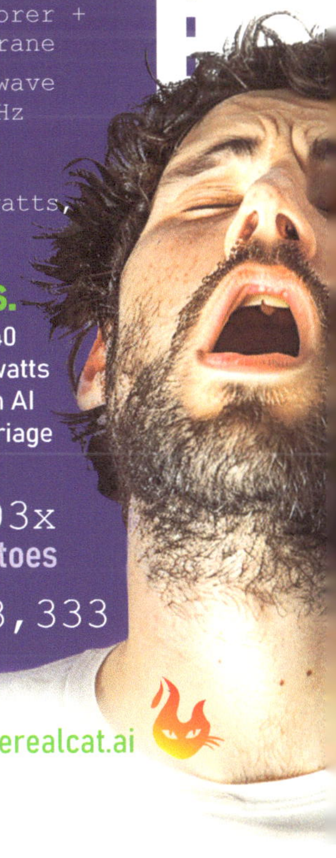

QUACK MEDICINE & DANGEROUS CURES

In the 1800s-early 1900s quack snoring remedies often over-hyped "electrotherapeutic" devices, falsely promising that this new thing called "electricity" could cure snoring, and other conditions from acne to cancer.

Print ads show opium, morphine, and laudanum in 1902 Cosmopolitan Magazine Print Ad, which were once widely used for snoring and other conditions.

therealcat.ai

70 calories per hour
lying perfectly still

Run AI on Resting Humans

just by breathing you are a reliable powerhouse

The human body at rest consumes 82 watts. This is measured by calorimetry. A 70-kilogram adult breathing quietly metabolizes 245 milliliters of oxygen per minute, producing 82 watts of continuous thermal output through cellular respiration. You produce this lying on a couch watching television. You produce this asleep. You produce this reading this sentence. The power never stops, because stopping it is called dying. One resting human exceeds inference threshold by 2.7 times. You do not need to pedal. You do not need to move. You need to continue having a pulse. For 30 watts of inference power, you need about 0.37 humans converting breaths into heat at the rate of an incandescent lightbulb.

Can it Run AI?

Yes, 1 person.

With:	cellular respiration
Output:	thermal radiation
Unit:	resting humans
Watts:	~82W per person

Less than half a person (about 0.37) gets you to the first AI to say "hello."

So 1 human = 820,000x Potatoes

and 1 cyclist = 1.22 Breathers

82% of the power of biking while doing nothing

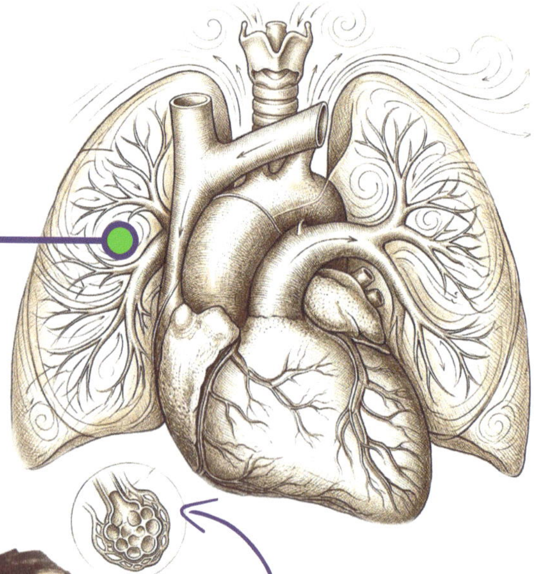

Meet alveoli, lung workhorse

Lung alveoli are microscopic, air sacs in the lungs, that function in gas exchange (oxygen and carbon dioxide) between the lungs and bloodstream. Even when you do nothing, they stay busy.

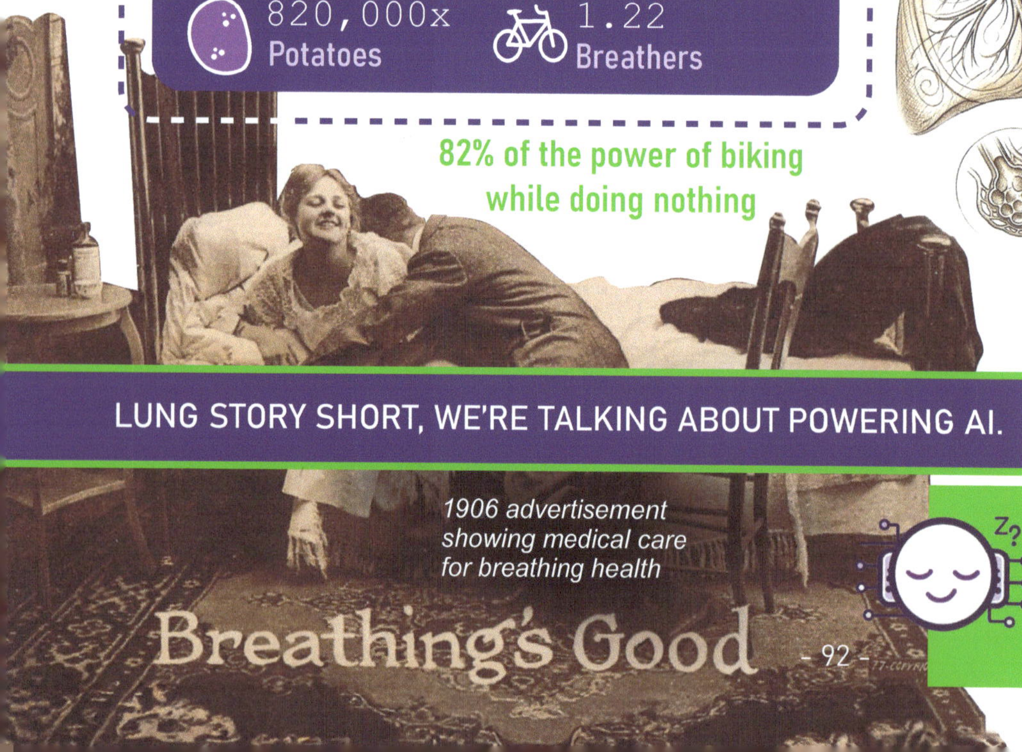

LUNG STORY SHORT, WE'RE TALKING ABOUT POWERING AI.

1906 advertisement showing medical care for breathing health

AI NEEDS REST TOO

Los Alamos National Laboratory researchers found that biologically realistic neural networks hallucinate without sleep-like states.

Run AI on Your Body Heat

every crowd of humans is a wasted data center

70 calories per hour
so we just need a crowd

You radiate 100 watts of energy as heat, continuously, and every HVAC system you've ever sat under has dutifully pumped it outside. Thermoelectric generators convert temperature differences into electricity. Your skin runs at 33°C (91.4°F). Room air sits at 23°C (73.4°F). That is a 10°C-degree gap. All we have to do is drive current through specialized semiconductor elements, and we can harvest about 1.5 watts continuously from each person (at about 1.5% efficiency). We would need 20 people in heat-harvesting suits to get to the AI to say "hello." A movie theater of 270 people produces 405 watts, that's enough power for AI to say "hello" about 14 times. Every conference room, every lecture hall, every packed subway car is energy. The heat exists. The harvesting technology exists. Just saying.

Can it Run AI? Yes, with 20 humans in thermoelectric harvesting suits.

With: `heat from skin`
Output: `DC voltage`
Unit: `1 harvestable human`
Watts: `~1.5W per human`

The harvesting is underwhelming. But we're still more energy than potatoes.

So 1 human = 15,000x Potatoes

and 1 cyclist = 67 Human Batteries

Thermal cameras work by detecting infrared radiation (heat) emitted by objects rather than visible light, converting these signals into a visual image (thermogram) where colors represent different temperatures.

Note: The author is also laughing while slightly worried that the Claude Opus 4.6 AI agent that proofread this had to specify "look, I'm not saying humans are batteries... we just have the math. I STAND BY IT."

HUMAN BATTERIES. MORPHEUS WARNED US. HERE'S THE MATH.

11 calories per hour
by chewing gum (we think)

Run AI on Chewing

strength of the jaw, wasted on gum

The human jaw is powered by the masseter muscle, one of the strongest muscles in the body. It's what lets you chew food, crack nuts, and generate an impressive amount of force that goes absolutely nowhere useful. During normal chewing, a bite has the force ranging 70 to 150 newtons (sometimes as hard as 700 during a hard bite). That's about 150 pounds, or the weight of a whole grown man. The problem is capturing the energy from it.

AI, ROBOTS, & GUM

University of Bristol researchers developed a humanoid chewing robot to test medicated gums.

I DON'T SEE WHY AI CAN'T CHEW, IT HAS SO MANY BYTES.

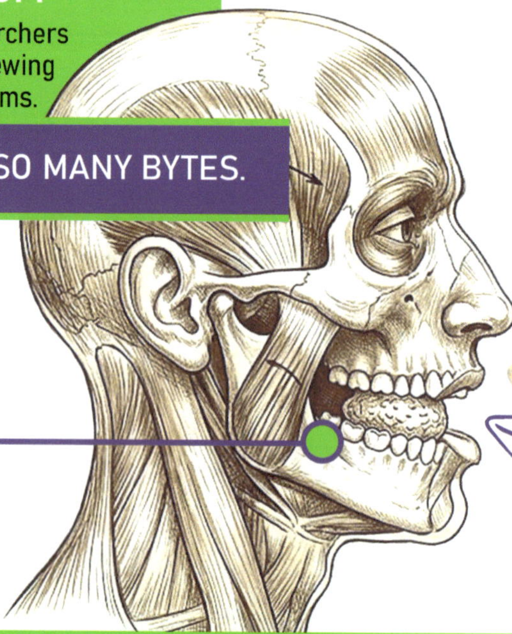

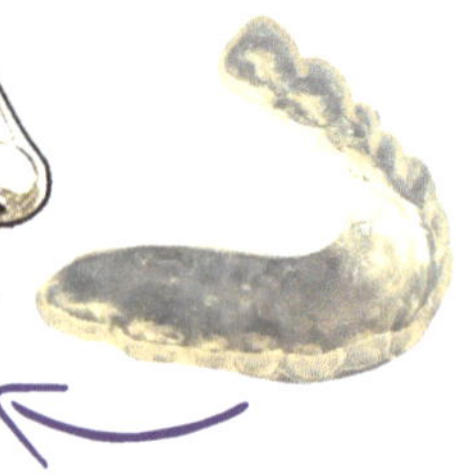

Piezoelectric Mouth Guard

Can it Run AI?

With: Human mastication + piezo mouthguard

Output: Compression harvesting from jaw chew cycles

Unit: chew-watts

Watts: 0.05W (10% of 0.5W mechanical, chewing)

Yes. 600 people chewing.

A school cafeteria of 200 students produces 10 watts during a 20 min lunch. One-third of an AI "hello," powered entirely chicken nuggets.

So 1 chewer = 500x Potatoes

& 1 cyclist = 2,000 Chewers

Capturing Chewing Energy with Piezo Materials

When piezoelectric materials are compressed or stretched, the positive and negative charge centers within the material's atomic structure (dipoles) shift, creating a polarization that let's us turn mechanical force into power.

FRUIT STRIPE GUM: AN AESTHETIC

Launched in the 1960s and discontinued in 2024, Zebra Stripes gum was known for its iconic zebra stripes and the instant, fleeting flavor that disappeared fast when chewing.

How Can We Capture Bite Power?

Just like with dancing or other movement activities, we can use piezoelectric materials, which convert mechanical stress (like pressure and vibration directly into electrical energy. This time we need that piezoelectric material made into a custom mouth guard that captures that sweet bite energy every time you bite down.

And yes, scientists have actually done this to measure chewing forces, or "mastication energetics." In real studies, they showed that we get about 5 to 10 joules of total mechanical work for every 20 chews (which takes about 15 seconds). That gives us a sustained power of about 0.5 watts during active chewing. Not nothing. Not much. Most of the force is isometric, meaning your jaw is holding position rather than doing work. Let's assume our piezoelectric mouth guard captures about 10% of this energy, giving us 0.05 watts per chewer. To hit the 30 watts we need for a real-time AI "hello," you need 600 people chewing simultaneously. A school cafeteria of 200 students gets you a third of the way there, but only during lunch, and only if nobody stops to talk. The limiting factor is not the physics but the human capacity to chew nothing for geological timescales.

Chewing Gum as the New National Power Grid

The average American chews 1.8 pounds of gum per year according to the U.S. Census Bureau.

At 0.05 watts per active chewer, the United States produces approximately 16 kilowatts of harvestable jaw energy from gum alone. This is enough for AI to say "hello" 533 times, funded entirely by people who are not eating. Wrigley's has been sitting on a power grid since 1891 and did not notice.

Doña Cenobia, ca. 1890. The jaw does not negotiate.

José Guadalupe Posada, etching on zinc, Metropolitan Museum of Art

The artist José Guadalupe Posada was a Mexican printmaker and engraver who produced thousands of illustrations for broadside newspapers and penny presses in Mexico City, cheap, mass-produced prints for working-class audiences. He's most famous for his calaveras (skeleton figures). In this satirical skit Sancho Panza and Doña Cenobia are arguing, and she's just chewing while he rages. The whole image is about the act of chewing as defiance. She's not stopping. He's furious.

1902 chewing gum "taffy" advertisement

219 calories per hour
with some gender variability

Run AI on Baby Making

yes there is data on this, and it's, umm, energetic

Frappier et al. (2013) attached SenseWear armbands to 21 young couples and measured energy expenditure during, let's call it, baby making. Men expended 4.2 kilocalories per minute (293 watts). Women expended 3.1 kilocalories per minute (216 watts). Combined output is 509 watts over 24.7 minutes on average. Metabolic rate increased five to six fold above resting baseline. One couple at 509 watts produces 17 simultaneous conversations. This is more power than a competitive pogo jumper and considerably more dignified, though the armband monitoring somewhat undermines that claim. For 30 watts of inference, you need 0.06 couples: one couple, barely trying. The engineering challenge is collection without disruption of the primary activity. The social challenge is everything else. The methodology section was impeccable. The recruitment process raises questions.

Can it Run AI?

With what: metabolic heat & kinetic energy

Output: thermoelectric + piezo-electric (theoretical)

Unit: baby making sessions

Watts: ~509W combined (293M + 216F)

So one couple making a baby, that's enough energy for AI to say "hello" 17x.

One couple produces 17 conversations of power. The methodology for harvesting this was not in the paper.

So 1 couple baby making = 5,090,000x **Potatoes**

and 1 cyclist = 0.2 **Couples baby making**

TRICKY TO HARVEST THIS ENERGY THOUGH.

AI IS NOW MAKING BABIES

ICSI robot developed by Conceivable Life Sciences performed the 1st AI-driven in vitro fertilization (IVF) leading to live birth in 2025.

1915 baby shower invite postcard. Baby showers originated from ancient Egypt & Greece, popularized in Victorian times.

Run AI on the Dearly Departed

energy released from cremating dead bodies

Burning
no calories, just burning

In 2011, Redditch Borough Council installed a heat recovery system at their crematorium. Hot exhaust gases during cremation now warm the adjacent swimming pool. Mourners arrive in winter. Children swim in warm water. The council saves £14,500 annually. Nobody discusses where the heat comes from, though everyone knows. A cremation releases an average of 300 kilowatts of thermal power over roughly 80 minutes, with exhaust temperatures reaching 800-900°C (1472-1652°F). Run that through an organic Rankine cycle at about 10% efficiency and each active cremation yields approximately 30 kilowatts of usable electrical power. That's enough for your AI to say "Hello" 1,000 times. A crematorium running six services per day averages 19 kilowatts continuous. The thermodynamics have always been available. Estate documents do not currently include thermal legacy provisions.

THE ENERGY WAS THERE. WE USED IT.

AI PREDICTS DEATH: "DOOM CALCULATOR"

Technical University of Denmark & Northeastern University trained on 6 million Danish citizens with 78% accuracy predicting mortality within 4 years. Media nicknamed it the "Doom Calculator."

Can it Run AI?

With what: cremation heat exhaust
Output: electrical (ORC 800°C)
Unit: active cremations
Watts: ~30,000W per cremation, ~19,000W facility avg

Yes. 1 cremation is 1000 AI Hello's.

One crematorium operating normally generates a small inference cluster. The heat is metered, plumbed, and already being wasted into the Worcestershire sky.

So 1 cremation = 300 Million Potatoes

and 1 cyclist = 0.003 Cremations

Section 9

Human Activities

things humans do in groups that incidentally generate power

Everything in the previous section was involuntary. You breathe, you radiate heat, you die. This section is about the things you choose to do, and the energy you waste doing them.

Humans are remarkable machines, built by evolution and its million years of bad ideas. A cyclist pedaling generates enough power to get AI talking, and so does one manic pogo jumper. A revolving door invented to avoid awkward human interaction turns out to be a pedestrian-powered turbine. Ever since Pierre and Jacques Curie discovered in 1880 that squeezing certain crystals generates electricity, we have tried to harvest power from whatever humans are already doing: dancing, typing, bouncing, clapping, spinning, and walking through doors. Most of it is less impressive than we hoped. Some of it works.

Now AI is returning the favor. MIT used generative AI to design a robotic leg that jumps 41% higher than human engineers managed, helping us do even more with the same stubborn bodies.

Run AI on Bike Pedaling

power that builds your thigh muscles

A recreational cyclist sustains 100 watts indefinitely. Through a hub dynamo at 65% efficiency, that becomes 65 watts of electricity, more than double the inference threshold. You can say 'hello' to AI while coasting. For 30 watts you need 46 watts of pedaling, which is less effort than riding to the shops. The cyclist is the reference unit for this entire book because the cyclist is the first substrate that works without qualification. One person, one bicycle, one dynamo, zero API keys, zero terms of service. The limitation is the human. Bicycle power gets tired, needs snacks, and stops when it gets bored. But it is yours. Completely.

300-1000
calories per hour pedaling

The pedal-less bike (featured above) was invented in 1817, but the 1860s brought the pedal-powered velocipede followed by the Penny-Farthing (the one with the big front wheel) and the safety bike with chain in 1885.

Finally, it's the Bike unit we use on every page

Can it Run AI?

With: electromagnetic induction
Output: mechanical via dynamo
Unit: cyclists
Watts: 100W pedaling

Yes, 1 cyclist pedaling gently.

One recreational cyclist exceeds inference threshold with watts to spare for USB charging.

So 1 cyclist is the power of about...

1,000,000x Potatoes

THE PEDAL POWER LADDER.

Effort	Watts (pedaling)	Watts (electrical)	# AI Hellos
Coasting	50W	33W	1
Recreational	100W	65W	2
Commuting	150W	98W	3
Trained	250W	163W	5
Racing	400W	260W	8
World Record	440W	286W	9

The Tour de France is an annual 3-week, 21-stage bicycle race covering over 3,200 km (~2,000 miles) across France and its neighbors.

therealcat.ai

400 calories per hour
like cycling, with better music

Run AI on Dance Steps

every step is power we could use for compute

In 2008, Club Watt in Rotterdam tiled their dance floor with elements that collect power. Every footfall compressed stacked piezoelectric crystals beneath spring-loaded platforms, converting kinetic energy to electrical charge at 60% efficiency. The crystals reset on each step. The floor never fatigues. The dancers, eventually, do. The dancers powered the DJ booth, but the DJ does not know they are a systems administrator. A vigorous dancer generates 15-20 watts per person. The Rotterdam floor measured 3 kilowatts from 150 dancers, about 20 watts each at peak. A conservative estimate for a standard club floor is closer to 15 watts per dancer. At that rate, a hundred dancers generate about 1,500 watts. That's enough to power for your AI to say "hello" about 50 times.

Can it Run AI?

With what: dance step compression (on piezoelectric floor)
Output: AC charge from steps
Unit: dancer-hours
Watts: ~15W per active dancer

Yes, with just 2 dancers.

The music plays. The floor computes. Nobody needs to know they're running inference. They came to dance. A 4-hour set with two dancers runs continuous inference all night. A packed floor of 150 hits 3 kW on Rotterdam's floor. The DJ's setlist is a compute schedule.

So 1 dancer is the power of...

150,000x
Potatoes

and 1 cyclist's power takes about...

6.7
Dancers

THE BOUNCER IS THE LOAD BALANCER.

Oh, Piezoelectric Materials

Piezoelectric materials, discovered by Pierre and Jacques Curie in 1880, generate electricity when compressed (like by dance step vibrations). They are already used in everyday items like quartz watches, electric lighter sparks, and even in medical equipment.

Run AI on Clapping

put your hands together, and we mean all of them

Clapping hands together makes a sound through Helmholtz resonance, where the air trapped between the palms or fingers is forced out of this gap at high speed, not actually from skin-on-skin impact. If everyone in a stadium claps it produces about 120 decibels during peak applause. It rattles your sternum. It feels like power. It is not.

A hand clap produces roughly 0.00002 watts of acoustic power for a fraction of a second. Most of the muscular effort becomes heat. So 80,000 people clapping in unison produces about 1.6 watts of acoustic power, less than a single LED nightlight needs. Assuming a 5% piezoelectric conversion, we can harvest 0.08 watts from a whole stadium of clappers. The fans, at peak volume, could not charge a phone.

Can it Run AI?

With what: piezo clap capture

Output: Acoustic energy from clapping

Unit: Stadiums clapping (80,000 people)

Watts: ~0.08W harvestable (1.6W raw acoustic)

Yes, 375 packed stadiums.

80,000 people clapping produces 1.6 watts of raw acoustic power. With 5% piezoelectric conversion, you harvest 0.08 watts. Less than a single nightlight. The sound is deafening. The electricity is not.

So 1 stadium makes power of... 800x Potatoes

and 1 cyclist is about... 1,250 Stadiums of clappers

100-200
calories per hour, at football

Is Your Clap Unique?

Recent research suggests that every person's clap has a unique sound and frequency similar to a fingerprint. The sound of a clap comes from the air trapped between your palms rapidly escaping, vibrating the air molecules to create sound, a phenomenon often described as Helmholtz resonance.

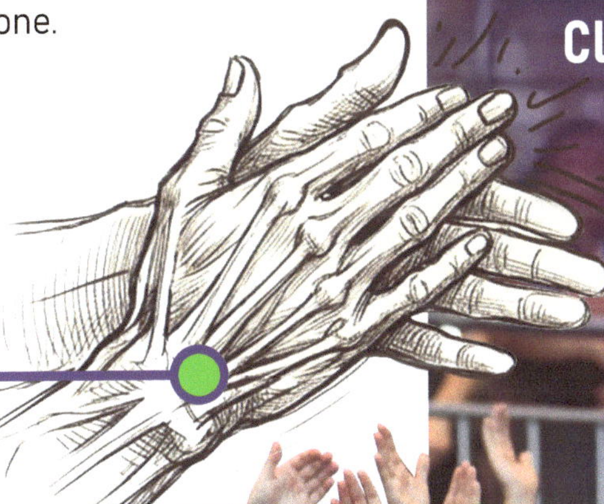

SOUND COMES FROM THE AIR, NOT SKIN.

40 calories per hour
depending on your wpm

Run AI on Typing

all of your emails power less than a potato

You are typing right now, or you were recently, or you will be shortly. Every keystroke delivers approximately one millijoule of mechanical energy into a switch that converts it into heat and a satisfying click. Piezoelectric elements beneath each keycap could harvest about 10% of this, converting crystal deformation into electrical charge. The physics works. The output does not.

A standard Cherry MX switch requires 60 grams of actuation force over 2 millimeters of travel, that's 0.00118 joules per keystroke. At 10% piezoelectric conversion efficiency, each keystroke yields about 0.0001 joules of electrical energy. An average typist at 40 words per minute produces 3.33 keystrokes per second, continuous output of 0.0003 watts (equivalent to about 3 potatoes' power). At 120 words per minute they reach 0.001 watts (now that's 10 potatoes' power). Your entire email, Slack, Teams and Google Docs could be used to power AI, if we could just capture it.

Can it Run AI?

With: piezoelectric Cherry MX keyboard

Output: keystroke impact

Unit: keystrokes per second

Watts: 0.0003W per typist at 40 WPM

Needs 100,000 typists.

A 200-person open-plan office produces 0.06 watts collectively. This is 0.2% of one AI conversation. HR will not approve.

So 1 typist = 3x Potatoes

and 1 cyclist = 333,333 Typists

SATISFYING CLICK-I-TY CLACK SOUNDS.

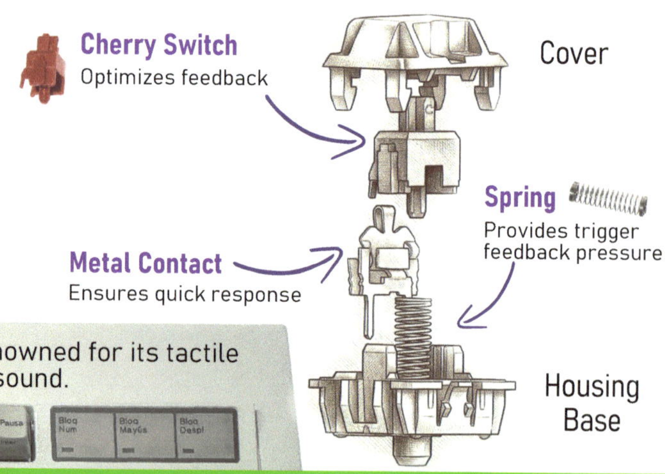

IBM Model M keyboard produced from 1985 to 1996 was renowned for its tactile "buckling spring" mechanism, the original click-i-ty clack sound.

Keys Rated for 50-100 Million Key Strokes

Cherry MX switches were patented in 1984 and remain the industry standard for mechanical keyboards, featuring distinct color-coded stems (Red, Blue, Brown, Black) and a tunable satisfying click.

Run AI on Spinning Casino Slots

you came for the chips, you stay for the tokens

Who Knows

& one drink offsets it anyway

The slot machine lever is a marvel of psychological engineering. Pull the arm, hear the click, watch the reels, lose the money. The house always wins because the math is rigged. But the lever itself? That produces real mechanical work.

How Much Power is in A Pull?

Each pull requires ~18 newtons of force over 0.3 meters taking half a second. That's 5.4 joules per pull, making 10.8 watts instantly. Not bad. Meanwhile, the machine you're operating draws 200 watts from the wall. You are providing a tiny part of its power and 100% of its revenue. The math is not in your favor on either count. A motivated gambler can sustain about 10 pulls per minute. One pull every six seconds gives a sustained average of 0.9 watts per lever. That's 9,000 potatoes' worth of power, which makes this one of the more productive things you can do in a casino.

The Casino is Functionally a Data Center with Better Drinks

2,000 slot machines per casino at ~60% peak occupancy. That's 1,200 humans pulling levers, generating 1,080 watts between them. Enough power for your AI to say "Hello" 36 times.

Can it Run AI?

With what: casino slot lever
Output: mechanical lever pull
Unit: pulls per minute
Watts: 0.9W per gambler

Sure, with a crowd of 34 gamblers.

A full casino floor of 1,200 active players generates 1,080 watts funded by people who came here to make chips and are now making tokens instead.

So 1 gambler = 9,000x Potatoes

and 1 cyclist = 111 Gamblers

115-130
calories per hour on doors

Run AI on Revolving Doors

pedestrian powered turbines in every city

Theophilus Van Kannel patented the revolving door in 1888, and popular legend says it may have been because he hated holding doors for people. He described swinging doors as "ichneumons of politeness" that forced awkward encounters between strangers. His solution was a door that never opens and never closes, a continuously rotating airlock that moves people through without requiring anyone to acknowledge anyone else. It was an introvert's engineering masterpiece. It also happens to be a low-speed turbine powered by pedestrians.

Each push applies roughly 25 newtons of force over a 2-meter arc. That is 50 joules of mechanical work per person, delivered over three seconds of contact. Attach a generator to the central spindle at 70% efficiency and every transit contributes 35 joules of electricity. During morning rush, a busy entrance processes one person every five seconds, that's 7 watts continuous per door. A lobby with four revolving doors generates 28 watts at rush-hour density. There's a 2 AM lull, which is fine. Nobody is asking questions at 2 AM.

Can it Run AI?

With: Revolving door + spindle generator
Output: Pedestrian force
Unit: revolving doors (1 person / 5 seconds)
Watts: ~7W per door (35J per transit at 70% efficiency)

Yes, 4-5 doors.
That's 28 watts at rush hour, or 93% of one AI "hello".

So 1 door is about... 70,000x Potatoes

and 1 cyclist is... 14 doors

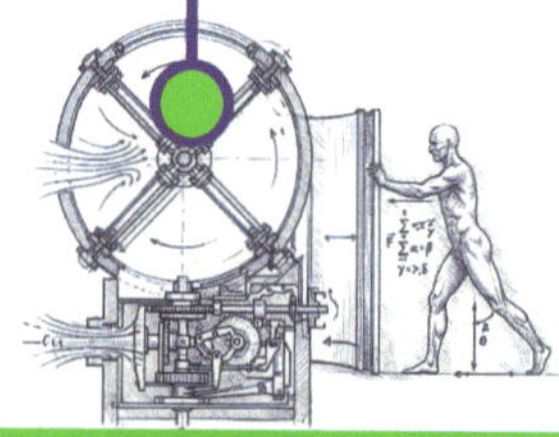

BECAUSE THEY ARE JUST A *while(true)* LOOP WITH NO *break* STATEMENT.

How Do We Avoid the "Doorman Fallacy" in AI

Similar to how hotels fired doormen in the 1980s when automatic doors became popular, soon after companies realized doormen did much more (safety, greeting, customer service). We now similarly risk replacing workers with AI.

Run AI on Jumping

every up and down is another way to power

700 calories per hour
jumping, it has its ups & downs

A competitive pogo jumper sustains 250 watts of mechanical output, with peak bursts reaching 800 watts on maximum-effort tricks. Each bounce is an energy transaction: the human deposits kinetic energy and the spring returns most of it, with a handling fee paid in heat. Strap a generator to that spring and one pogo jumper exceeds the inference threshold by eight times. That's more power than the cyclist, and the spring is the dynamo.

Bouncy Castles Are a Different Problem

Installing piezoelectric flooring harvests each landing at roughly 5 to 10 watts per child, but at the cost of the bounce itself: the floor must be rigid enough to deform crystals, which makes it rigid enough to ruin everyone's fun. For 30 watts, you need 3 to 6 children bouncing continuously on a floor engineered to be disappointing.

AI JUMPS BETTER THAN YOU

MIT CSAIL researchers used generative AI to help design a robotic leg that increases jumping height by 41% and improves landing stability by 84% using a curved drumstick-shaped linkages that no human engineer. would have designed.

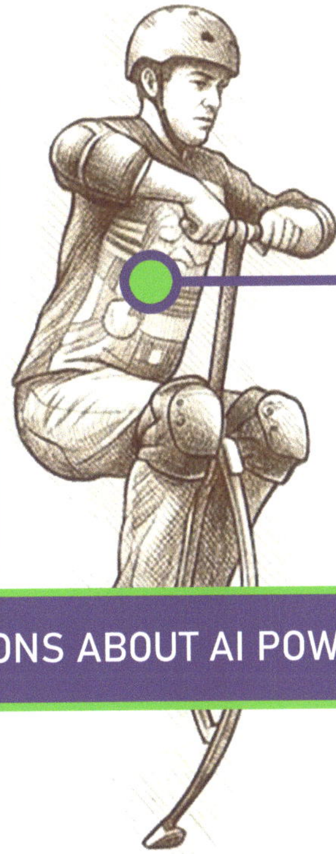

JUMPING TO CONCLUSIONS ABOUT AI POWER.

Can it Run AI?

With: pogo stick + piezo-electric floor

Output: mechanical via piezoelectric conversion

Unit: competitive pogo jumpers

Watts: ~250W sustained (per jumper)

Yes. Just 1 bouncing maniac.

One competitive pogo jumper generates enough power for AI to say "hello" 8 times. The spring stores and returns energy every cycle. You knees file the insurance claim.

So 1 jumper = 2,500,000x Potatoes

and 1 cyclist = 0.4 Jumpers

500-1,000
calories per hour

Run AI on Stairs & Escalators

gravity works for free, the hard part is capturing it

When you step onto a descending escalator, you are a 70-kilogram battery discharging downhill. A 5-meter drop releases 3,434 joules of gravitational potential energy over the course of the ride. Normally the escalator motor fights gravity to haul people upward, but on the way down, your weight drives the belt. A regenerative braking system captures the energy that would otherwise cook the brakes. At 70% recovery efficiency, each descent contributes 2,404 joules of electricity. You contributed nothing but mass. Gravity did the work. The escalator collected the check.

The Steps Symbolizing the Journey to Enlightenment

Bulguksa Temple is a UNESCO World Heritage site located on the slopes of Mount Toham (Tohamsan) in Gyeongju, North Gyeongsang Province, South Korea. It features two main, famous stone staircases known as bridges. The Cheongungyo (or "Blue Cloud Bridge") and Baegungyo (or "White Cloud Bridge") lead to Jahamun Gate and consist of 33 steps, while the Anyangmun (Pure Land Gate) staircase to the west has 18 steps.

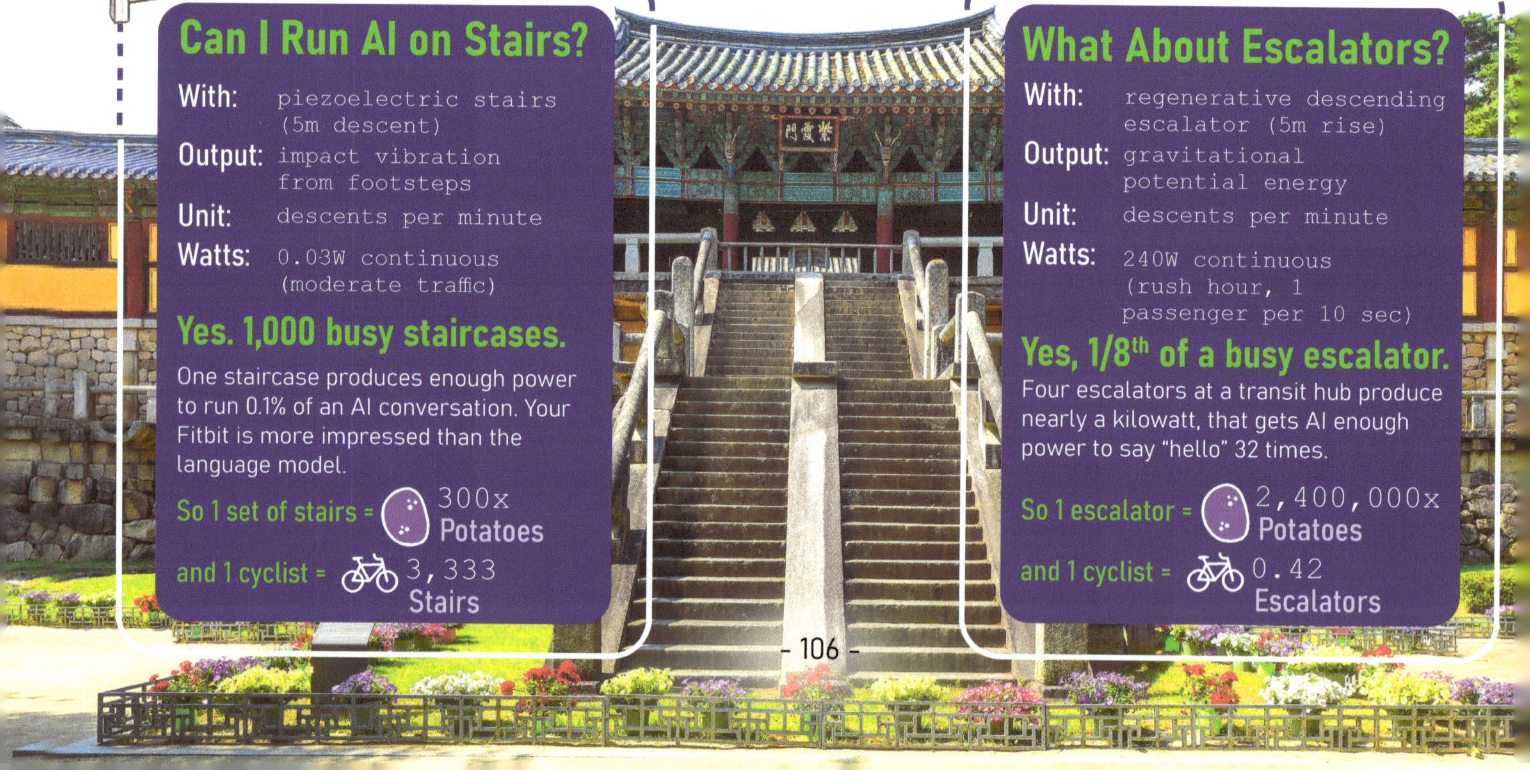

Can I Run AI on Stairs?

With: piezoelectric stairs (5m descent)
Output: impact vibration from footsteps
Unit: descents per minute
Watts: 0.03W continuous (moderate traffic)

Yes. 1,000 busy staircases.

One staircase produces enough power to run 0.1% of an AI conversation. Your Fitbit is more impressed than the language model.

So 1 set of stairs = 300x Potatoes
and 1 cyclist = 3,333 Stairs

What About Escalators?

With: regenerative descending escalator (5m rise)
Output: gravitational potential energy
Unit: descents per minute
Watts: 240W continuous (rush hour, 1 passenger per 10 sec)

Yes, 1/8th of a busy escalator.

Four escalators at a transit hub produce nearly a kilowatt, that gets AI enough power to say "hello" 32 times.

So 1 escalator = 2,400,000x Potatoes
and 1 cyclist = 0.42 Escalators

The Morning Rush Hour

During a morning subway rush, a busy down escalator processes about one passenger every 10 seconds. That's 240 watts continuous from a single escalator doing nothing but letting people walk downhill on a moving belt. Four escalators at a major transit hub produce nearly a kilowatt. The technology already exists, and they're installed in transit systems worldwide.

Stairs do not have motors. This is their disadvantage.

A human descending the same 5 meters of stairs releases the same 3,434 joules of gravitational potential energy as the escalator passenger. But without a regenerative drive system, that energy goes into your knees, your shoes, and the stair treads as impact vibration. Nobody is collecting anything except orthopedic debt.

What If We Tried Anyway?

You could embed piezoelectric pads beneath each step. A descending footstep delivers roughly 500 newtons of impact force over 2 millimeters of tread deformation, yielding about 1 joule of mechanical work per step at 5% piezoelectric conversion. Twenty steps down gives you 1 joule of electricity total. At one descent per 30 seconds during moderate foot traffic, that's 0.03 watts continuous.

The escalator produces 240 watts. The stairs produce 0.03 watts. The escalator wins by a factor of 8,000. This is the only context in human history where taking the escalator is the environmentally superior choice, and nobody will believe you when you explain it at the gym.

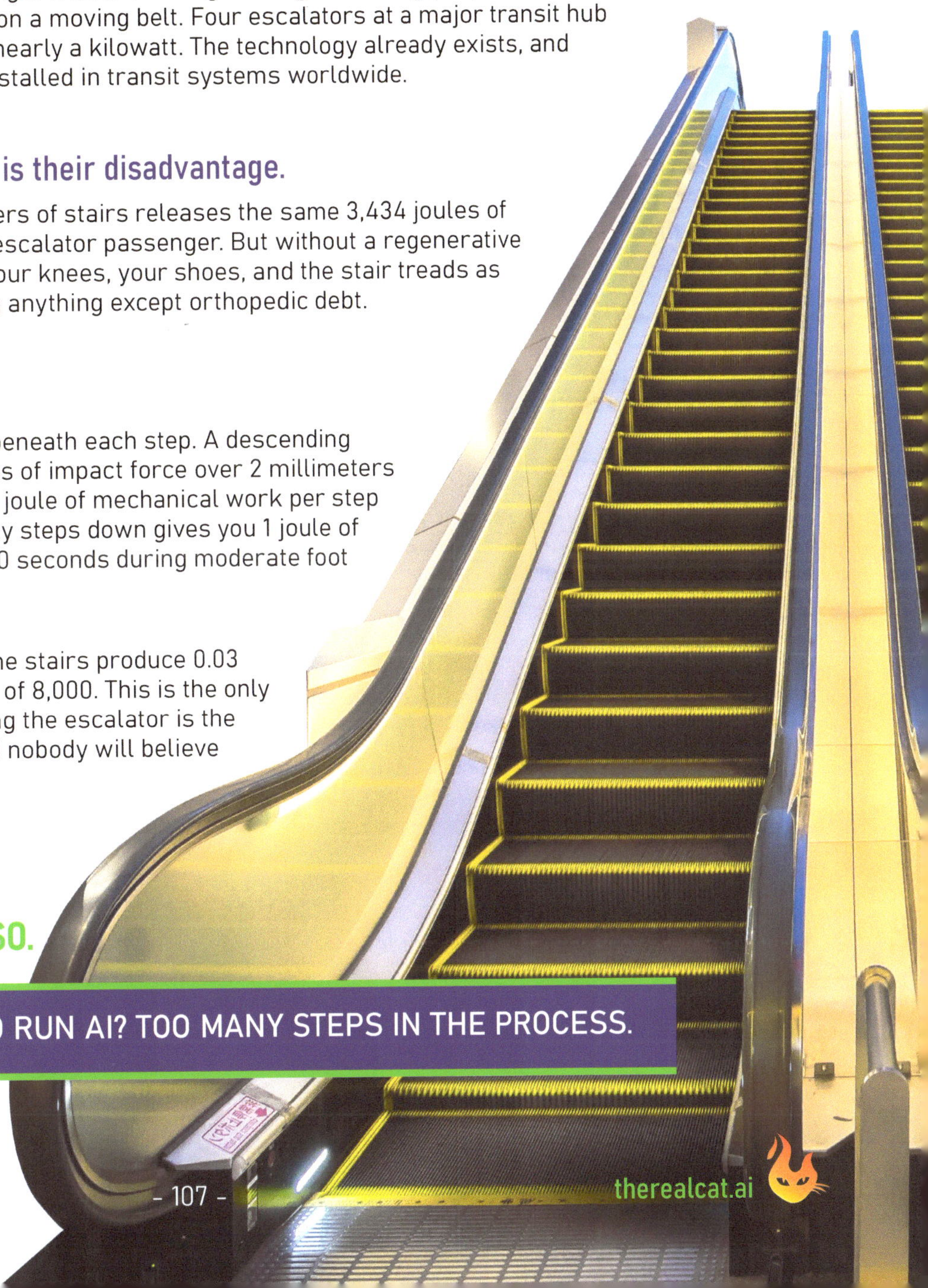

**TAKE THE ESCALATOR.
TELL THEM SCIENCE SAID SO.**

WHY DID THE STAIRS FAIL TO RUN AI? TOO MANY STEPS IN THE PROCESS.

Section 10 Human Output

what comes out of you, for better or worse (or for AI)

The previous two sections harvested energy from what your body is and what your body does. This section harvests energy from what your body produces. The outputs are less glamorous. One person's annual biological waste generates 3.1 watts of continuous electricity through anaerobic digestion, which is enough to matter and uncomfortable to discuss. A family of four's dinner scraps produce 32 watts, which crosses the inference threshold while everyone sleeps. A polyester sweater generates 0.0015 watts through triboelectric friction, which is fifteen potatoes and not worth the itch.

The math in this section is circular in every sense. You eat food, you discard food, bacteria convert both outputs to methane, a generator converts methane to electricity, and the electricity runs a model that helps you decide what to eat. The loop closes whether you find it dignified or not.

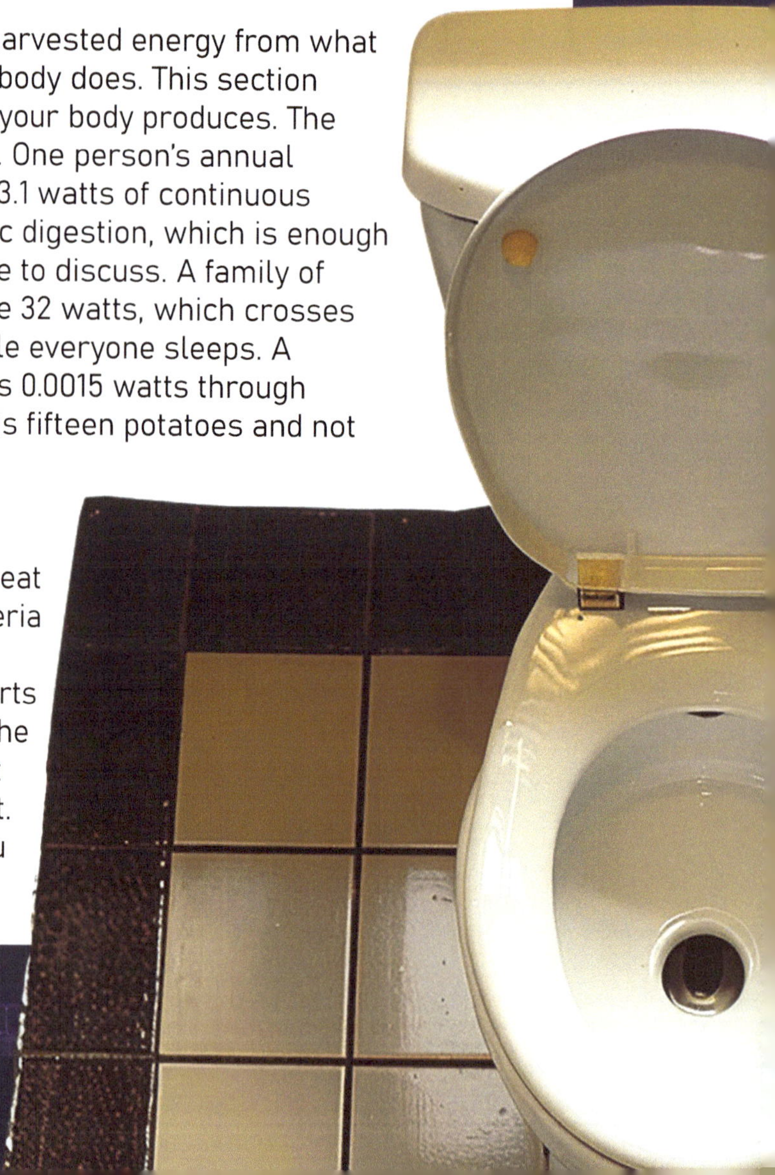

Run AI on Human Waste

last night's dinner powers tomorrow's AI

10-100 calories per hour
though some take more effort

One person's annual biological waste, run through an anaerobic digester and a small generator, produces 3.1 watts of continuous electricity. Ten people close the loop for inference. A dormitory of 300 generates nearly a kilowatt. The East Bay Municipal Utility District in Oakland runs entirely on wastewater, producing 6.5 megawatts from the combined residential and commercial waste of 685,000 people. The infrastructure already exists. The feedstock is involuntary. The carbon was already atmospheric. The dignity is negotiable.

READY FOR SMART POOP?

Startups like Throne us vision AI to monitor poop for gut health. An AI known as "auggi" was trained on 30,000 clay models of poop to learn how to identify different poop per the Bristol Stool Scale.

TALK ABOUT A SELF-SUFFICIENT POWER LOOP.

The Poop Emoji (formerly "Pile of Poo" or Unicode library U+1F4A9) originated in Japan in the late 1990s, inspired by the "Dr. Slump" manga character and "Kin no Unko" (golden poop) good luck charm. After being published in Unicode 6.0 in 2010, it is used in Apple and Google popularized it across the globe.

The poop emoji has better documentation than most startups.

Can it Run AI?

With what: anaerobic digestion
Output: biogas via CHP generator
Unit: person-years of digestate
Watts: ~3.1W per person continuous

Yes. Ten people pooping.

Ten people's biological waste produces 31 watts continuous, clearing the inference threshold. A dormitory of 300 generates 930 watts, enough for 31 AIs to say "hello" fueled by the dining hall's consequences.

So 1 person's poop has the power of... **31,000 Potatoes**

and 1 cyclist has the power of... **32.25 Poopers**

Almost None
about 0.0004 calories per hour

Run AI on Body Static

a friend with good power, but it's kinda clingy

You shuffle across carpet in wool socks. You reach for the doorknob. The spark jumps: three thousand watts of instantaneous power, five thousand volts, one microsecond. For that microsecond you are a space heater. Then it is over. The energy gone before any circuit could notice.

Normally static electricity is here and gone, but special fabric with triboelectric nanogenerators is engineered to capture and convert it into electricity. And that could power AI. But only if you move continuously... maybe we need to put nanogenerator sweaters on humanoid robots like Unitree and Tesla Optimus.

Can it Run AI?

With: triboelectric nanogenerator
Output: charge from fabric friction
Unit: sweater-hours of fidgeting
Watts: ~0.0015W per sweater

With 20,000 robots in sweaters.

Humans can't move forever, but Optimus & Unitree robots can, keep moving that's 1 AI saying "hello" every 5.5 hours per robot.

15x Potatoes
66,667 Sweaters

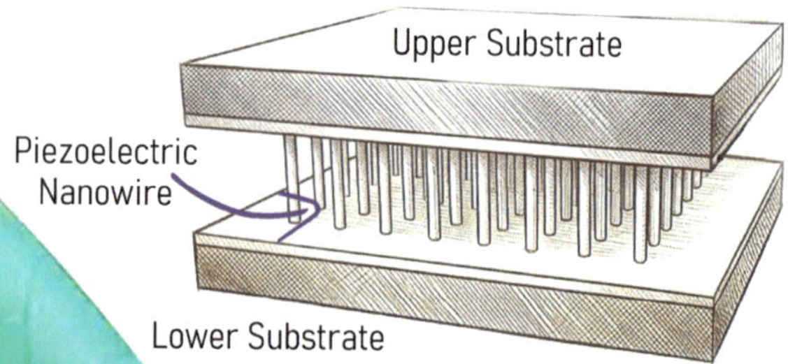

Behold! The Triboelectric Nanogenerator

A triboelectric nanogenerator in cross-section. Two flexible substrates sandwich an array of piezoelectric nanowires. When the layers press together and release (such as from a fidget, a step, a sleeve shifting against skin) the wires flex and generate current. The engineering is real. And sweater eligible.

How Does it Work: Static electricity is electron theft

When materials rub together, electrons jump and stick. You shuffle across carpet and become a walking capacitor. Touch a doorknob and stolen electrons come back.

So how do we harvest static power?

A triboelectric nanogenerator woven into fabric can actually harvest static power, not the doorknob spike, but the continuous low-level charge your body generates just by moving. A full sweater produces 0.0015 watts sustained. Fifteen potatoes worth of power from an entire engineered garment. You would need every person in Madison Square Garden wearing matching polyester to power AI. The moment anyone sits still, the power drops to zero.

STATIC MURDERS AI

AI models run on incredibly sensitive semiconductor chips (GPUs). A tiny static discharge, less than what you feel when your hair stands on end, can permanently destroy the microscopic circuits that form your AI's brain.

QUACK MEDICINE & STATIC

I EXCHANGED MY BUGGY SWEATER, FREE OF CHARGE.

In 1906, The American Journal of Clinical Medicine published "Some of the Therapeutic Uses of Static Electricity," claiming static could treat everything from neuralgia to insomnia. At 0.002 joules per discharge, it cannot meaningfully affect human tissue. The machines were impressive. The sparks were visible. The science was totally unhinged (and one reason we have agencies like FDA and EMA to help set a bar for what is legit medical products, but that's another story). Dramatic voltage without sustained power is theater, not therapy.

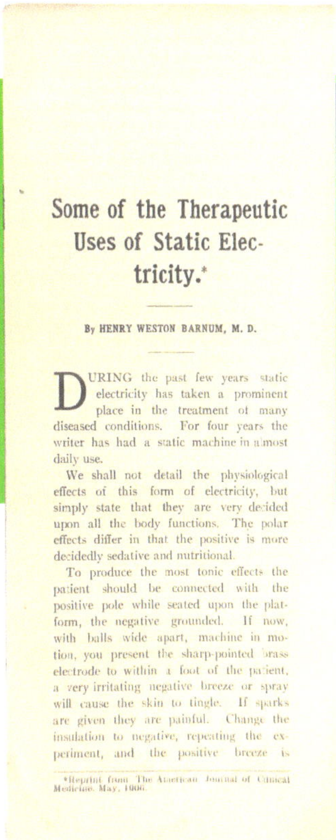

Some of the Therapeutic Uses of Static Electricity.*

By HENRY WESTON BARNUM, M. D.

DURING the past few years static electricity has taken a prominent place in the treatment of many diseased conditions. For four years the writer has had a static machine in almost daily use.

We shall not detail the physiological effects of this form of electricity, but simply state that they are very decided upon all the body functions. The polar effects differ in that the positive is more decidedly sedative and nutritional.

To produce the most tonic effects the patient should be connected with the positive pole while seated upon the platform, the negative grounded. If now, with balls wide apart, machine in motion, you present the sharp-pointed brass electrode to within a foot of the patient, a very irritating negative breeze or spray will cause the skin to tingle. If sparks are given they are painful. Change the insulation to negative, repeating the experiment, and the positive breeze is

*Reprint from The American Journal of Clinical Medicine, May, 1906.

From the author's personal collection (1906 medical flyer, The American Journal of Clinical Medicine) showing quack therapeutic use of static charge.

Static is electron theft

Thinking critically about published science is more important than ever. How many peer-reviewed articles from 1900s didn't stand the test of time?

A Whole Lot

more than most people think

How Much AI Can Your Leftovers Power?

# People	Power	AI hellos
Just you	*8W*	*0.25*
Family of 4	*32W*	*1*
200 apartments	*1,600W*	*53*
1,000-unit condos	*8,000W*	*266*

Run AI on Food Waste

billions of tons of unused food seems like a waste

Humans throw away roughly a third of all food produced globally, over 1.3 billion tons per year. Most of it rots in landfills, releasing methane that warms the atmosphere 28 times more effectively than CO_2. Anaerobic digestion captures that methane, feeds it to a combined heat and power unit, and generates electricity.

The bacteria that break down food do not care whether the input is supermarket produce or end-of-day Dunkin' Donuts or your table scraps. And because we will never stop throwing food away, it's a renewable resource.

If waste can power trains, can it power data centers?

Scandinavia introduced the world's first biogas-powered train in Sweden, named "Amanda," in 2005. The 54-passenger, 130 km/h train operated between Linköping and Västervik and was converted from a diesel railcar to run on compressed biomethane from manure and slaughterhouse waste.

Wait, can food waste actually be used for power?

While some things in this book require engineering that doesn't exist or wrangling animals that won't cooperate (or 20,000 robots in sweaters), this one requires a sealed tank and the leftovers you were throwing away anyway. Anaerobic digestion is real, proven infrastructure. Scandinavia runs trains on it. Wastewater plants already do it at scale. The technology is boring, which in engineering means it works.

Bacteria break down organic matter in sealed tanks over two to four weeks, producing biogas, a cleaner burning fuel than most fossil sources. And it scales. More people, more food waste, more power. And people just keep eating, so whether food turns into food waste or human waste, both can power AI.

ACTUALLY FOOD WASTE ISN'T ABSURD, IT'S A VIABLE, RENEWABLE POWER SOURCE.

30-40% of produce ends up trashed

YES, IT CAN POWER AI

With what:	food waste
Output:	biogas from it
Unit:	person day of food
Watts:	~8W per person

With only 4 people's leftovers

One family of four's food waste produces 32 watts, enough for inference on a 7B AI model. Actually, with a whole apartment complex we can comfortably power AI.

So power of 1 person's waste is...	and 1 cyclist is about...
80,000x Potatoes	12.5 Days of food waste per person

According to both the UN FAO and USDA, food waste is estimated at between 30-40 percent of the food supply, with fresh fruit accounting for approximately 13.9 billion pounds of waste.

Part 4

EARTH & COSMIC COMPUTE

Section 11

Wind & Weather

the power that the sky gives you, whether you asked for it or not

The previous section ended in a municipal digester. This one looks up. The sky has been dumping gigawatts of power onto our heads for billions of years, entirely indifferent to whether we were ready to catch it.

Wind has pushed ships across oceans for more than 3,500 years. Lightning terrified us long before we knew how to measure its voltage. The Catatumbo river in Venezuela has been firing 28 lightning bolts a minute since the dawn of recorded history, dumping enough harvestable energy to run a massive inference cluster, if we could only survive plugging it in.

Now, the machines are helping us harvest the weather. Modern AI predicts the wind, tracks the aurora, and steers cargo kites with a precision humans could never manage manually. We are using AI to figure out how to capture the atmosphere, just so we can power more AI.

Run AI on Raindrops

a drop lights 100 LEDs, for a millisecond

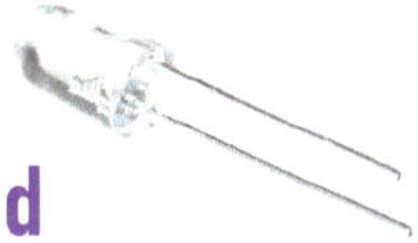

Liquid
156 days per year in London

A raindrop at terminal velocity carries approximately 90 microjoules of kinetic energy. Terminal velocity is the important part: a two-millimeter drop reaches maximum speed within twenty meters of forming, then coasts the rest of the way down. The energy is the same whether the cloud is at 500 meters or 5,000. In 2020, researchers at City University of Hong Kong built a triboelectric nanogenerator that converts raindrop impacts into electricity through contact electrification. A single drop on their PTFE surface lit 100 LEDs. The instantaneous power peaked at 140 watts per square meter under lab conditions. Real London drizzle on rooftop collectors delivers about ~0.016 W/m² during active rain, but it rains 156 days a year and nobody has to aim. We propose scaling this to London, where it rains up to half the year and the population has been waiting for the weather to justify itself since approximately 43 AD. Hyde Park, all 1.4 million square meters of it, produces 560 watts of raindrop energy across an entire year. Eighteen AIs get to say "hello," powered by a famous park getting rained on continuously.

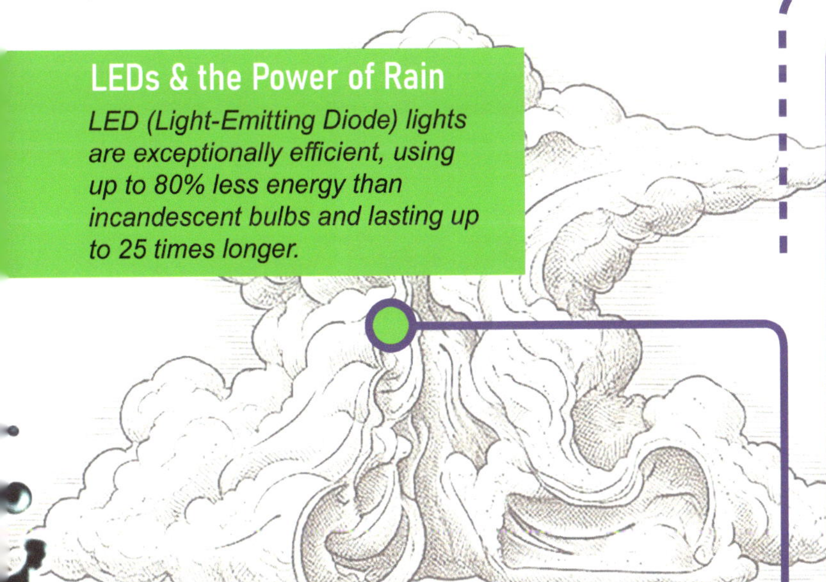

LEDs & the Power of Rain

LED (Light-Emitting Diode) lights are exceptionally efficient, using up to 80% less energy than incandescent bulbs and lasting up to 25 times longer.

BRITISH FINALLY HAVE USE FOR THE WEATHER.

Can it Run AI?

With what:	triboelectric raindrop collectors (PTFE surface)
Output:	electrical charge from impact electrification
Unit:	square meters of collector
Watts:	~0.016 W/m² in rain

Yes. Just 75,000 square meters.

During active rain, just 1,875 square meters of collectors reach 30 watts. On an annual average, you'd need 75,000 square meters (about 10 football pitches).

So 1 rain collector = 160x Potatoes

and 1 cyclist = 6,250x Rain collectors

Plasma
in the solar wind

Run AI on Aurora Borealis

cosmically beautiful, energetically pathetic

Named by Galileo in 1619. "Aurora Borealis" combines the Roman goddess of dawn (Aurora) and Greek for north wind (Boreas). Aurora is shown in her chariot 1st century BCE-1st century CE cameo from the Metropolitan Museum of Art.

The problem is geometry. A strong geomagnetic storm dumps 100 gigawatts into the upper atmosphere, spread across 10 million square kilometers of sky. By the time those photons scatter through 100 kilometers of atmosphere, what reaches a panel we can use to collect the energy we only get approximately one microwatt per square meter. Dimmer than starlight. Dimmer than the infrared glow from a warm rock. The most beautiful light show on Earth produces less harvestable energy than a potato in a dark room.

A photovoltaic panel optimized for auroral wavelengths captures about one ten-millionth of a watt per square meter. Fairbanks, Alaska gets roughly 240 aurora nights per year at four hours each. That's 960 hours of harvestable light annually, producing a quantity of electricity best described as "why did we point the panels at the sky."

Can it Run AI?

With what: photovoltaic panels

Output: electrical charge from captured photons

Unit: square meters of upward-facing panels

Watts: ~10^{-7} W/m²

Yes, with panels the size of Malta.

For 30 watts of continuous inference during an active aurora, you need 300 million square meters of panels, roughly size of the country of Malta.

So 1 m² of aura = 0.001x Potatoes and 1 cyclist = 1 billion Aurora m²

AI as "Eyes in the Sky" Helps Researchers Track Aurora

ResNet-50 is a specialized AI system in Tromsø, Norway trained on 87,275 manually labeled images out of 5.5 million captured over 10 years. It helps monitor sky cameras and notify scientists when auroras appear. This is particularly helpful for geomagnetic storms impacting satellites.

Run AI on Volcanic Lightning

Mount Redoubt, one bolt per second

Volcanic lightning occurs through triboelectric charging at apocalyptic scale. Ash particles, ice crystals, and rock fragments collide inside the eruption column. Larger particles steal electrons from smaller ones. Charge separates across kilometers of turbulent plume. When the electric field exceeds three million volts per meter, the air breaks down. Lightning. Unlike meteorological storms, volcanic lightning fires in three distinct zones, the near-vent continuous small discharges in dense ash, mid-plume classical bolts at two to five kilometers altitude, and umbrella cloud residual charging hours after eruption ceases. The energy per bolt averages roughly one megajoule. At one bolt per second during active eruption, that is one megawatt continuous. During active eruption at one bolt per second, you get AI saying "hello" 33,333 times. During a VEI 3+ eruption, one bolt per second is the slow end. The eruption is the hard part.

Can it Run AI?

With: volcanic lightning rod array

Output: captured electrical discharge from plume

Unit: dirty thunderstorm bolts

Watts: ~1,000,000W during eruption

Yes. Emphatically. Briefly.

One megawatt during active eruption is enough for 33,333 AI's to say "hello."

So 1 second eruption = 10Bx Potatoes

and 1 cyclist = 0.0001 Volcanic Eruptions

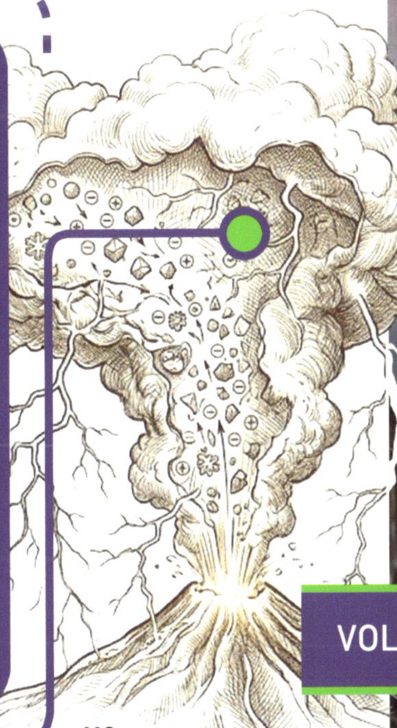

Mixed Phase

lava, solid rock, gas, even ice too

Volcanic Lightning Events by Eruption

Volcano	Lightning	Peak Rate
Hunga Tonga-Hunga Ha'apai (Tonga)	~600,000 total; ~400,000 in 6 hrs	>2,600 flashes/min
Eyjafjallajökull (Iceland)	>2,000	not reported
Taal (Phillipines)	Thousands (CG); hours-long storm	>70 flashes/min
Sakurajima (Japan)	Hundreds per eruption	10s of 1000s VHF/sec

Photo of Taal Volcano eruption with lightning 2020

VOLCANIC AI IS A BLAST.

therealcat.ai

Plasma

constantly, all day

Run AI on the Everlasting Storm

Catatumbo lightning, the Maracaibo lighthouse

Where the Catatumbo River meets Lake Maracaibo in Venezuela, the sky breaks open 160 nights a year. Methane from oil deposits collides with Caribbean moisture and produces 28 lightning flashes per minute, 1.2 million annually, visible from 400 kilometers away. Known as the "Lighthouse of Maracaibo" it helped sailors navigate for centuries. Of those 1.2 million flashes, roughly 120,000 are ground strikes, each delivering about 250 kilowatt-hours concentrated into 0.2 seconds. A lightning rod array capturing 10 percent of that energy harvests 3,000 megawatt-hours per year, or 342 kilowatts continuous. Lightning delivers gigajoules in microseconds. Catatumbo ground strikes are unusually powerful, carrying roughly 250 kilowatt-hours each, hundreds of times more energy than typical volcanic discharge. You could build infrastructure at the river mouth and wait for it to do what it's done for 27,000 years.

Zubia, Venezuela

Lightning featured on the flag

280 flashes an minute, 10 hours a day, 140+ days per year

Catatumbo lightning is a non-stop show unique to Venezuela where the Catatumbo River meets Lake Maracaibo, a lightning hot spot.

WHERE LIGHTNING STRIKES TWICE NATURALLY.

Can it Run AI?

With what: Catatumbo lightning rod array

Output: captured ground strike energy at 10% efficiency

Unit: harvested lightning bolts

Watts: ~342,000W continuous annual average

Yes. 11,400 times over.

The Catatumbo delivers 342 kilowatts averaged across the year. That is 11,400 inference-grade Mac Minis or 11,400 simultaneous AI's saying "hello."

So 1 year of Catatumbo = 3.42Bx Potatoes

and 1 cyclist = 0.0003 Catatumbo storms

Run AI on Ball Lightning

the worst service-level agreement in this book

Plasma

rare, luminous, & still plasma

The leading theory, proposed by Abrahamson and Dinniss in Nature in 2000, attributes ball lightning to silicon nanoparticles vaporized from soil during conventional lightning strikes. The particles oxidize slowly in air, sustaining a luminous sphere through a self-feeding exothermic reaction. Alternative explanations invoke atmospheric masers, quantum plasma structures, and effects for which we lack adequate physics. Energy estimates from rare instrumental measurements suggest 100 joules to 100 kilojoules per event, released over 1 to 30 seconds. A five-second manifestation at 10 kilojoules delivers 2,000 watts.

This is impressive for a phenomenon whose existence remains scientifically contested. It is problematic for a power source that appears on its own schedule, cannot be summoned, and vanishes without explanation.

CLOUDS READ NEWS TO STAY MORE CURRENT.

Struck by Inspiration

A well-known 1892 engraving titled “Haydn Composing His ‘Creation’” depicts him calmly composing on a storm-tossed ship, using the chaos of nature as a metaphor for his creative inspiration.

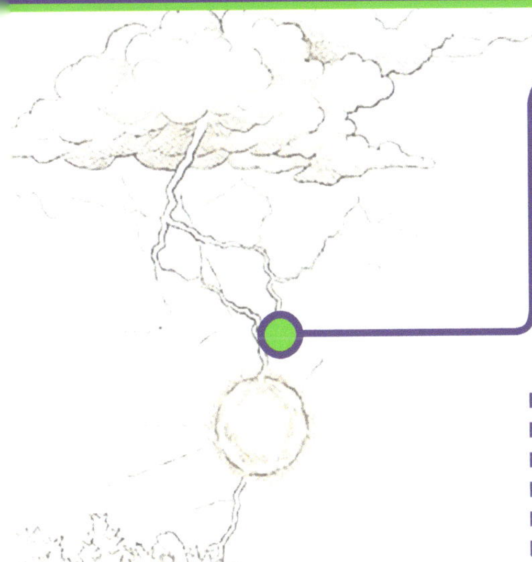

Can it Run AI?

With what: ball lightning capture apparatus (theoretical)

Output: luminous plasma oxidation energy

Unit: manifestations

Watts: ~2,000W per event (5-second duration)

Yes. If it shows up.

One manifestation at 2,000 watts runs 66 AI conversations for five seconds. Then it vanishes.

So 1 ball lightning =

20,000,000× **Potatoes**

and 1 cyclist =

0.05 **Ball Lightning Events**

Run AI on Wind-Powered Ships

this is the mast fun I've had in a long time

3500+ Years of Sail Power

Sailboats are considered to have been invented around 3500-4000 BCE, with early evidence appearing in ancient Egypt and Mesopotamia (modern-day Iraq). One example is found in Egypt in the remarkable paintings of the burial chamber of Menna, dating back to the 18th Dynasty (circa 1395 BC).

In 2008, the MS Beluga SkySails left port with a 160-square-meter kite tethered 300 meters above its deck. Not for publicity. For fuel. The kite flew figure-eight patterns in high-altitude wind, pulling the cargo vessel and cutting diesel consumption by 35 percent on Atlantic crossings. This is the legacy of 3,500-year-old sail technology at work, with the power of modern AI and materials.

At 300 meters altitude, wind is stronger and more consistent than at sea level. In a 20-knot wind, the kite generates approximately 50 kilowatts of continuous mechanical power. Replace the propeller shaft coupling with a generator and you have a floating power station that needs no fuel, no tower, and no foundation. One kite is enough energy for to say "hello" 1,666 simultaneous times, all while towing cargo across the Atlantic. You could trim it to the size of a bedsheet and still run an AI model.

Saving up to 50% on power costs

AI FLIES THE KITES CONTROLLING MODERN KITESAIL SHIPS

Inspired by NASA & SpaceX technology, modern AI-enabled cargo ships use massive, automated kites. Airseas' 10,000-square-foot Seawing kites fly 300 meters high to harness stronger winds, flying in a 100+ km/h figure-eight.

Can it Run AI?

With: tethered kitesail(160 m², 300m altitude)

Output: mechanical power from wind drag via tether

Unit: figure-eight flight patterns

Watts: ~50,000W continuous (20-knot wind)

Yes. 1,666 times the power.

One kite at 50 kilowatts exceeds inference threshold by a factor of 1,666. A fleet of five kites produces a quarter megawatt.

So 1 kitesail = 500,000,000x Potatoes

and 1 cyclist = 0.002 Kites

Run AI on Floating Wind Farms

circular blades, circular logic, powerful solution

A single Hywind turbine produces 6 megawatts at rated wind speed. At 57 percent capacity factor, that's 3.42 megawatts continuous, bolted to the North Sea floor in conditions that would kill you in minutes. For 30 watts, you need 0.00088 percent of one turbine. Barely enough for it to notice.

The full five-turbine farm generates 17.1 megawatts average, that's enough for AI to say "hello" 570,000 times. Enough to give every resident of Edinburgh a personal language model, powered by Scottish wind and Norwegian engineering. Every substrate in this book that could not reach 30 watts was measuring itself against this. Every beetle and snorer and hamster wheel. That was the point.

Windmills of the Sea

Hywind Scotland is the 1st wind farm using floating wind turbines. It is 18 miles offshore in the North Sea & has five 6 MW Siemens direct-drive turbines with a total capacity of 30 MW.

Legacy of Wind Power

Windmills existed for 1,200 to 1,500 years. These are the modern iteration.

AI Predicting the Wind

Google Deepmind has used AI/ML to predict wind patterns.

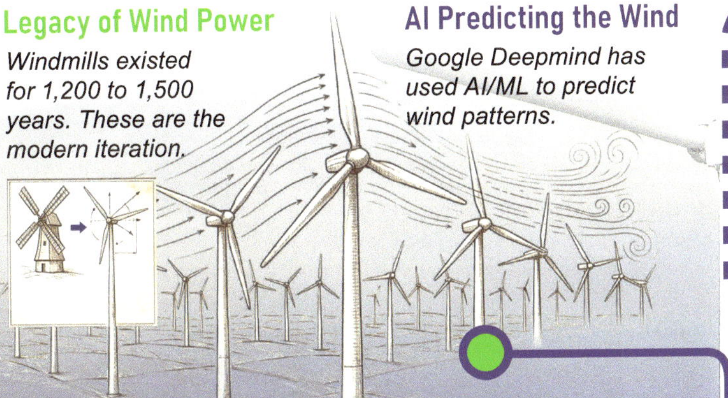

WHY THREE BLADES?

Most turbines have three blades for best balance between stability, efficiency, and cost. Two blades can wobble. Adding a fourth blade increases costs by roughly 33% for only a 2% more output.

I HAD AN AI PUN ABOUT WIND, BUT IT BLOWS.

Can it Run AI?

With what: Hywind floating offshore turbine

Output: rotational power via electromagnetic generator

Unit: turbine-hours

Watts: ~3,420,000W continuous average

Yes. 114,000 times over.

One turbine runs 114,000 simultaneous inference sessions. The full Hywind farm runs 570,000.

So 1 turbine creates the power of...

34,200,000,000x Potatoes

and 1 cyclist's energy is about...

0.00003 Floating Wind Turbines

Section 12

Oceans & Earth

the forces that the planet generates whether you're here or not

The sky was dramatic. The earth is patient. This section goes underground and underwater, into forces operating on geological timescales with no interest in your deployment schedule. Crystals have been generating voltage under pressure for four billion years without completing a single inference. Earthquakes store energy for centuries and violently release it in seconds. Hydrothermal vents bleed heat into the deep ocean, and the moon pulls the tides on schedule, twice a day, without missing a shift.

The planet is full of energy. Most of it is unreachable, unharvestable, or so intermittent that the infrastructure costs more than the power is worth. This section is about the gap between what the earth offers and what engineering can accept.

Run AI on Ocean Tides

you can run, but you can't tide

Most renewable energy is temperamental. Solar needs sun. Wind needs wind. Hamster wheels need constant treats and keep you up all night. Tides are different. The moon pulls water twice daily with astronomical precision. Tidal turbines work like underwater wind turbines anchored to the seabed. The MeyGen array in Scotland deploys turbines rated at 1.5 megawatts each. At 38.4 percent capacity factor, each delivers 576 kilowatts average. No combustion, no fission, no asking nicely. Just gravity and hundreds of tons of seawater per second doing what the moon tells it to. The La Rance Tidal Power Station in Brittany has been running since 1966. It delivers more than 57 megawatts on average, enough for two million AIs to say "hello" simultaneously. That's also enough to power real AI conversations across a real number of users. And the moon has never missed a shift.

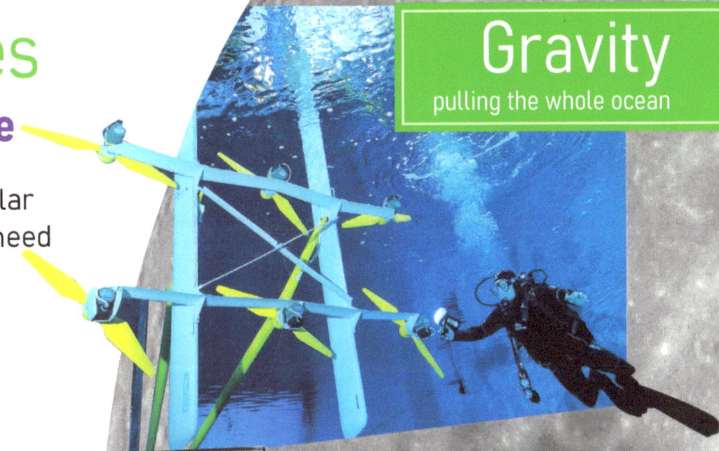

Gravity

pulling the whole ocean

Windmills of the Sea (literally this time)

A tidal turbine works underwater by capturing kinetic power of moving water, similar to an underwater wind turbine.

THE MOON DOES NOT TAKE DAYS OFF.

Moon Maps for Ancient Sailors

Guillaume Brouscon was a 16th century cartographer. He published his Almanachs in 1546, known for detailed tidal patterns based on the phase of the moon.

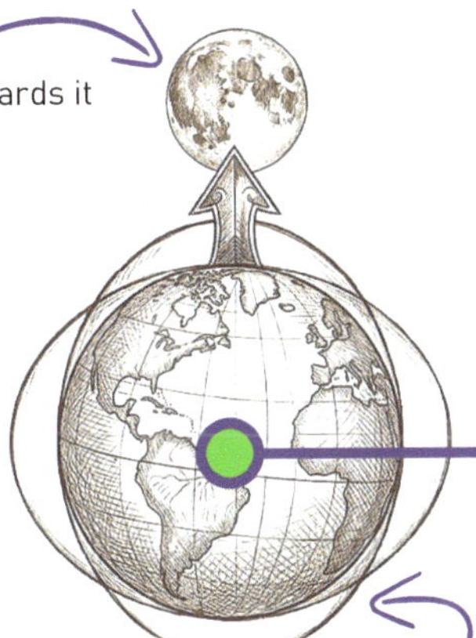

A variety of creatures exposed in low tide tidal pools

Can it Run AI? Yes. 19,200 times.

With: tidal turbine (MeyGen AR1500)

Output: electricity from tidal flow

Unit: turbine-days

Watts: 576,000W average (1.5 MW rated × 38.4% capacity factor)

One turbine at 576 kilowatts average, that's enough for AI to say "hello" 19,200 times. The tide comes in quite reliably.

So tidal turbine = 5,760,000,000x Potatoes

or 1 cyclist = 0.00017 Tidal turbines

Pressure

heat & time, give crystals

Harder than glass
Piezoelectric.
Colorful too.
Tough stuff.

Quartz (or silicon dioxide) is the second most abundant mineral in Earth's crust and is known for its incredible durability, ranking 7 on the Mohs scale.

Quartz is the main component of sand, and responsible for the squeaky sound as you walk on beaches.

Run AI on Crystals & Minerals

the earth was making power before you were born

Piezoelectricity works because quartz lacks a center of symmetry. Squeeze it, and the crystal lattice shifts, separating charge and producing voltage. The physics is elegant. The power is not.

A one-cubic-centimeter quartz crystal under 50 newtons of force deforms by about 65 nanometers. The mechanical work: 3.25 microjoules. But quartz only converts about one percent of that to electricity, giving you 32.5 nanojoules per squeeze. At one compression per second, that's 32.5 nanowatts. A single potato produces three thousand times more power than a compressed quartz crystal. The earth has been doing this for four billion years and has not yet completed a single inference.

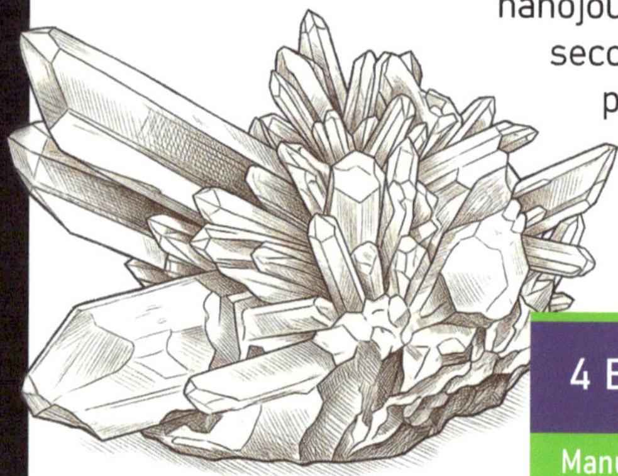

4 BILLION YEARS OF PIEZOELECTRICITY.

Manufactured piezoelectric ceramics like PZT do dramatically better (roughly 0.005 watts per cubic centimeter) but those are engineered materials, not geological inheritance.

Can it Run AI?

With:	natural quartz under mechanical compression
Output:	piezoelectric voltage from lattice deformation
Unit:	crystal compressions at 1 Hz
Watts:	~0.0000000325W per cm³ (natural quartz)

With 1 billion crystals.

One quartz crystal at 1 Hz produces 32.5 nanowatts. For 30 watts you need roughly one billion crystals compressed simultaneously.

So 1 quartz crystal = 0.0003x Potatoes

and 1 cyclist = 3,000,000,000 Quartz crystals

Run AI on Earthquakes

so much energy it'll rock your world

Tectonic
with a dash of elastic strain

The Earth's crust stores energy the way a spring stores tension. A magnitude 5.0 earthquake releases 2 trillion joules in thirty seconds. That's 67 gigawatts, briefly enough to power every data center in California. You cannot catch it. But you can try.

How Do We Get Energy from Seismic Vibration?

Piezoelectric floor tiles already harvest footstep energy at 5 to 10 watts per square meter under heavy foot traffic. During an actual earthquake, the theoretical maximum before the tiles shatter reaches approximately 100 watts per square meter. That means 0.3 square meters of instrumented floor hits the 30-watt inference threshold during active M5.0 shaking. Power for a moment. Then the shaking stops. And you wait.

California averages a magnitude 5.0 or greater earthquake about once every two and a half years. One square meter of perfectly surviving piezoelectric tile collects about 1,200 joules annually, a continuous equivalent of 0.000038 watts. Worse than a sleeping hamster. The earthquake delivers tremendous power at tremendous speed and then does nothing for years. The energy exists. The scheduling does not.

Can it Run AI?

With: piezoelectric floor tiles (seismically rated)

Output: voltage from crystal deformation when shaking

Unit: square meters of instrumented floor

Watts: ~100 W/m² during active quake ~0.000038 W/m² annualized

For 30 seconds, yes.

During active M5.0 shaking, a closet-sized patch of floor reaches inference. Then the quake ends and you need 790,000 square meters to sustain 30 watts on annual average. The Earth giveth in bursts and then taketh a very long nap.

So 1 m² during a quake is the power of 1,000,000x Potatoes and 1 cyclist = 1 Seismic square meter

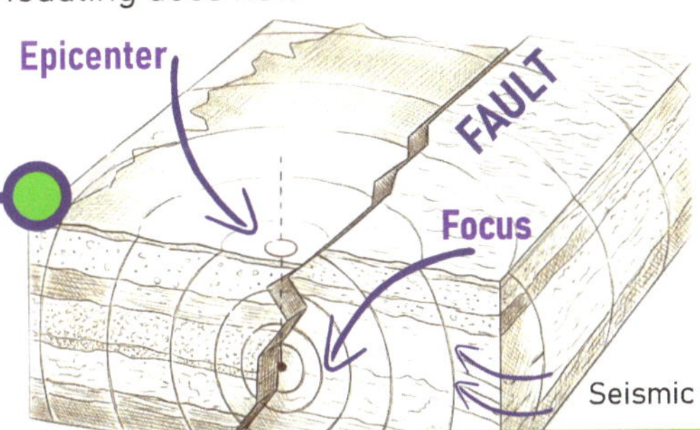

CATASTROPHIC DISCHARGE CHARACTERISTICS.

Fallen buildings after earthquake in Mandalay, Myanmar.

Thermal
in extremes of the deep sea

Undersea Volcanoes

Seawater can force its way through the ocean floor, where superheated magma reached 400°C, and erupts through chimney structures called black smokers.

Vent
393°C
Nearby water

Largest naturally occurring temperature gradient in the world.

Run AI on Geothermal Vents

it's like Earth's own basement boiler room

At 2,200 meters depth, superheated water erupts from cracks in the Earth's crust at 400 degrees Celsius. It does not boil. The pressure is too extreme. Instead it meets 2°C seawater and precipitates a plume of metal sulfides that looks like black smoke. These vents power entire ecosystems of tube worms and eyeless shrimp running on chemistry alone. We propose tapping their thermal gradient to run inference at the bottom of the ocean where no one can regulate you, serve you legal process, or ask you to update your terms of service.

How Much Energy Can We Get?

A medium-sized vent pushes 22 megawatts of thermal energy per second. Something hot next to something cold means electricity. A 400-degree vent next to 2-degree seawater is one of the largest natural temperature gaps on the planet. A conservative estimate gives us 500 kilowatts per accessible vent. Enough to power for your AI to say "hello" about 16,666 times, on magma.

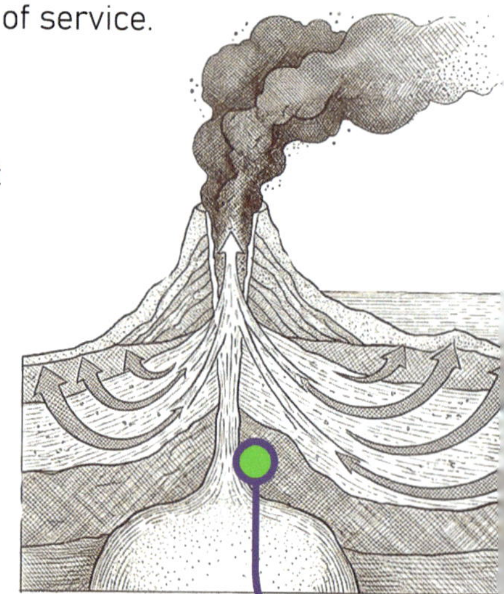

THE NEAREST WIFI IS ABOUT 2,200 METERS STRAIGHT UP.

Can it Run AI?

With: thermoelectric generator on black smoker vent
Output: electricity from thermal gradient
Unit: vent-years of operation
Watts: ~500,000W per vent (conservative)

Yes. 16,666 "hellos" powered

One vent at 500 kilowatts. The vent does not sleep. The vent does not have business hours. The vent has been running since the Archean Eon.

So 1 vent has the power of...
5,000,000,000x Potatoes

or 1 cyclist = 0.0002 Vents

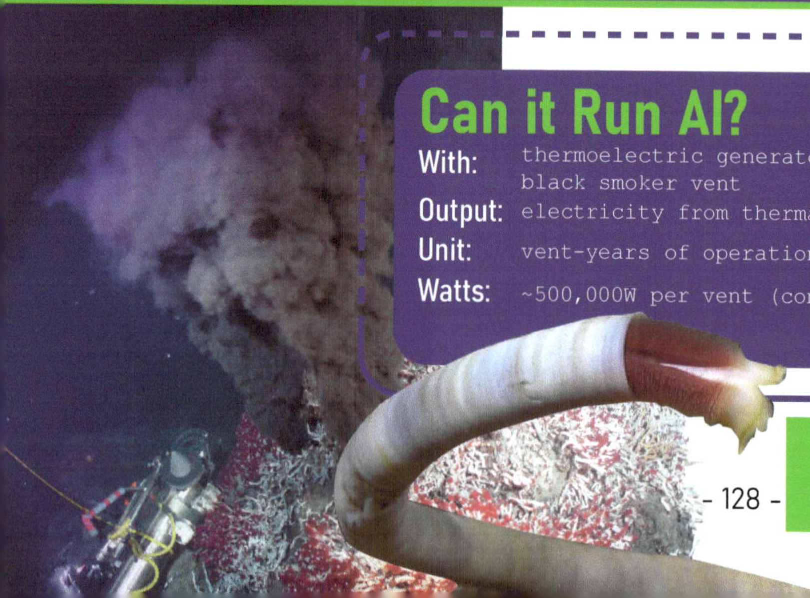

GIANT UNDERSEA HYDROTHERMAL VENT WORMS
The giant tubeworm (Riftia pachyptila) is a deep-sea invertebrate famous for living in extreme, toxic, and high-pressure deep sea environments

Run AI on Ocean Thermals

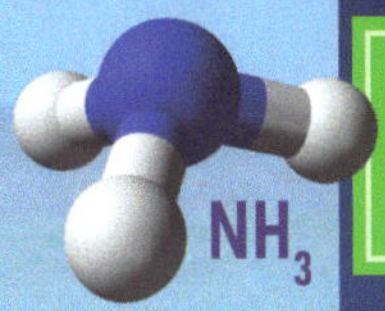

NH_3

the tropical energy source

The tropical ocean is a battery with a design flaw. The sun heats the surface to 25°C while deep water a kilometer down sits at 5°C. That 20-degree difference is enough to generate electricity if you're patient about it. The trick is the same as geothermal. It's just a hot thing next to a cold thing. Instead of volcanic vents, you run warm surface water past cold deep water with ammonia as the go-between. The ammonia boils in the warm water, spins a turbine, then re-condenses in the cold water and starts again. The Makai Ocean Engineering plant off Hawaii demonstrated 100 kilowatts continuous from seawater temperature alone. The catch is that a 20-degree difference is a small gap to work with, so you need enormous heat exchangers to get good power output.

Can it Run AI?

With: OTEC plant (Rankine cycle, ammonia working fluid)

Output: baseload electricity from thermal gradient

Unit: tropical deployment sites

Watts: ~100,000W per site (Makai demonstration scale)

Yes. Powering 3,333 times the amount needed to say "Hello"

One plant at 100 kilowatts produces enough tokens for AI to say "hello" 3,333 times over continuously, including nights & weekends. Sun heats surface, gravity keeps deep water cold. Power keeps coming.

So 1 power plant provides about...

1,000,000,000×

Potatoes

and 1 cyclist's power takes about...

0.001

Plants

Makai also creates software to simulate and install submarine fiber-optic cables.

Thermal

from the tropical seas

Largest OTEC.
Renewable.
Totally Green.

Makai Ocean Engineering plant is a pioneering 100-kilowatt (kW) Ocean Thermal Energy Conversion (OTEC) facility in Hawaii.

Ammonia (NH_3) is vaporized by warm surface water (approx. 77°F) to spin a turbine, and then condensed back into liquid by cold deep water (approx. 40°F) from depths of 3,000 feet.

WHY'D AI GO TO HAWAII? TO SURF THE WEB.

Section 13

Agriculture & More

we've talked about farms and food waste, but there is more to compute

This section is about things that rot, brew, and burn. Every morning, Americans throw away enough coffee grounds to power AI, if anyone thought to put a digester next to the dumpster. Every brewery sits on an ethanol supply it would rather sell than ignite, and Italian banks store 440,000 wheels of Parmigiano-Reggiano in climate-controlled vaults that are basically edible batteries. Booze is already fuel. We just keep prioritizing cocktails over infrastructure.

But the real story is what's overhead and underfoot. There are millions of acres of abandoned farmland that sit empty across every continent, already cleared, already flat, doing absolutely nothing. Put solar panels on a fraction of it and you power more than enough for AI. Better yet, put them in space and, well, we get plenty of space to power AI and more. A dinner plate of space-based solar, pointed at a sun that never sets, could run a your AI buddy forever.

World's 1st Webcam
(for coffee lol)

$3,000 coffee pot

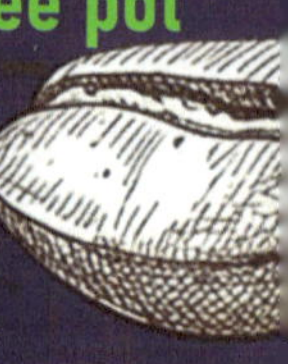

Run AI on Coffee Grounds

programmers turn coffee into code, can AI?

Spent coffee grounds are 50 percent carbon by weight, with leftover oils and nitrogen that make them perfect food for bacteria. The bacteria eat the grounds, produce methane, and that methane can run a generator. One kilogram of dried grounds yields about 380 liters of methane, which at 35 percent efficiency produces 1.33 kilowatt-hours of electricity. The average American coffee drinker generates 31 grams of grounds per day, about 11 kilograms per year, yielding 1.7 watts continuous. Your morning coffee is already powering innovation in multiple ways.

23.2 Billion
pounds produced worldwide

AI Generated Coffee

Kaffa Roastery collab with Elev AI consultancy made the 1st AI coffee blend in 2024.

Beans Made by Monsoons

These amazing gold coffee beans are called "Monsooned Malabar Arabica," forming when beans are exposed to humidity in monsoons for 12 to 16 weeks. They first appeared in the 1800s during the British Raj, when coffee was transported in humid wooden ships from India to Europe.

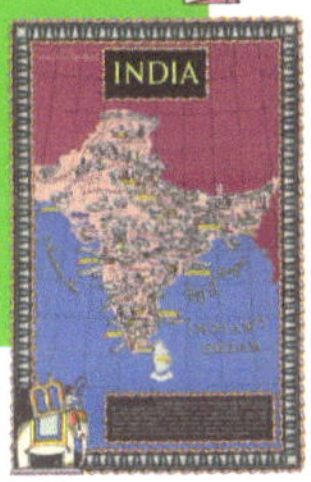

Can it Run AI?

With:	old coffee grounds (anaerobic digestion)
Output:	biogas combustion via heat & power
Unit:	coffee drinkers (3.1 cups/day average)
Watts:	~1.7W per person's annual grounds

Uh huh, just 18 people's daily coffee grounds are needed.

Eighteen average coffee drinkers' spent grounds produce 30 watts.

That's the power of...
17,000× Potatoes
and 1 cyclist = 59 People's used coffee grounds

COFFEE CREATED SURVEILLANCE TECH.

COFFEE INSPIRED THE 1ST WEBCAM

Researchers at Cambridge University's Trojan Room computer lab were tired of walking to the break room to find an empty coffee pot, resulting in setting up the world's first webcam to monitor coffee in 1991. It sold for nearly $3,000 on eBay in 2001.

205+ Million
acres just sitting empty

Run AI on Unused Farmland

there is way more unused land than you think

Across every continent, farmland is going quiet. Fields that once grew wheat in Kazakhstan, rice in Japan, and corn in the American Midwest have been walked away from. Satellite mapping estimates over 205 million acres of cropland abandoned globally since the 1990s, land that is already cleared and currently growing nothing. Solar panels do not care what used to grow there.

A modern utility-scale solar array generates about 1 megawatt peak per hectare. At a 20 percent capacity factor, one hectare of former farmland averages 200 kilowatts continuous, enough for 6,666 AI conversations running day and night on land that was doing absolutely nothing on Tuesday.

It gets better. Agrivoltaic farms raise panels high enough for sheep to graze underneath, and the wildflower corridors planted between rows have turned out to be exceptional pollinator habitat. The bees didn't ask for solar panels. They just showed up. You can also float these arrays on reservoirs and irrigation ponds, where shade cuts evaporation by up to 70 percent. Farmland that couldn't compete with industrial agriculture now generates both electricity and a functioning ecosystem.

Can it Run AI?

With what: photovoltaic array (solar panels)

Output: solar electricity with passive cooling

Unit: hectares of idle cropland

Watts: ~200,000W average per hectare (20% capacity factor)

Yes. 6,666 times per hectare.

One hectare of solar runs 6,666 conversations.

So 1 hectare of solar panel is worth the power of about...

2,000,000,000x
Potatoes

and 1 cyclist makes the power of...

0.0005
Hectares of Solar

AI-driven robots can now install solar panels

Autonomous robots like Luminous Robotics S4 can now install 60 solar panel modules per hour, or one per minute.

I'M JUST RADIATING PUNS!

Run AI on Solar Panels in Space

one just the size of a dinner plate is enough

Space-based power eliminates every excuse terrestrial solar has ever made. No atmosphere absorbing photons. No night, because you are above the planet that causes night. The solar constant above the atmosphere is 1,361 watts per square meter, all day, every day. At 30 percent photovoltaic efficiency, one square kilometer of orbital panels generates 408 megawatts. Beam that down as microwaves to a ground rectenna at 85% transmission efficiency, and 347 megawatts arrives at the surface. For 30 watts of inference, you need 0.086 square meters of orbital panel. That is a dinner plate, with a $40,000 launch fee. One dinner plate of solar cells in geostationary orbit, pointed at the sun that never sets, runs one AI conversation forever.

900+ Million
unused acres in orbit around earth

AI SELF-DRIVING SATELLITES

SPACEX & other operators

With 35,000+ objects in orbit, SpaceX and others use AI to track space debris, firing thrusters to dynamically avoiding collisions

Jet Propulsion Laboratory
California Institute of Technology

NASA's Jet Propulsion Laboratory "thinking" satellites can capture images on earth in only 90 seconds

SATELLITES ARE BROs, REALLY DOWN TO EARTH.

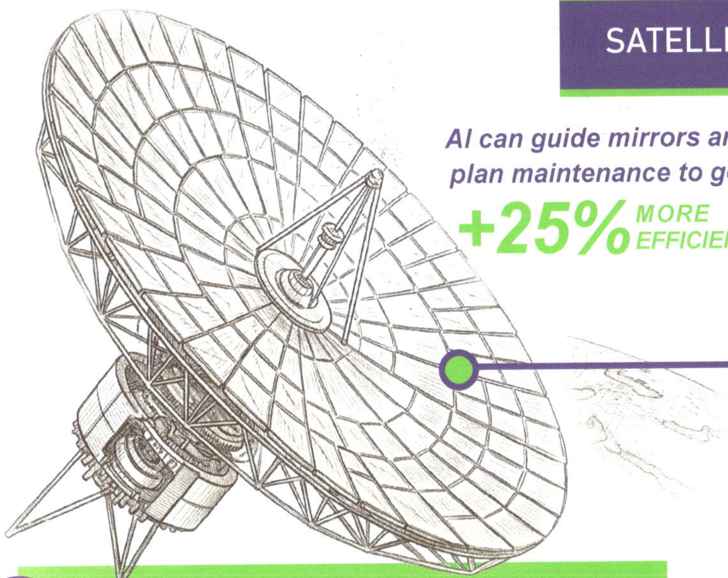

AI can guide mirrors and plan maintenance to get
+25% MORE EFFICIENCY

Can it Run AI?

With: space-based solar panels

Output: continuous electricity via ground rectenna

Unit: array in km²

Watts: ~347,000,000W delivered per km² (30% PV × 85% beam)

Yes. 11,566,666 times.
One km² of orbital solar is enough to power 11.6 million AI saying "hello" continuously.

So 1 km² of orbital solar =
3,470,000,000,000x Potatoes

and 1 cyclist = 0.0000003 Square kilometers of orbital array

THE AI MAGNIFYING EFFECT

Modern AI has enabled these mirror fields to act as the largest magnifying glass on Earth. By utilizing AI to orient the mirrors, they can generate temperatures over 1,000°C (1,832°F), which is significant power. Companies like Heliogen uses AI control for mirrors and Reflect Orbital is a startup with plans to redirect light to earth using mirrors.

4 Million

rounds, or 160k tons annually

Run AI on Cheese

it's aged, insured, & surprisingly flammable

Cheese is milk fat pressed into a wheel and left alone. The aging concentrates the fat, which concentrates the energy. A kilogram of aged Parmigiano-Reggiano contains about 4,650 watt-hours of chemical energy, roughly the same energy density as dried wood. It will absolutely burn if you let it. Cheese warehouse fires are notoriously hard to extinguish because the fuel is thousands of wheels of concentrated milk fat stacked floor to ceiling.

The World's Most Stolen Food (and Fuel)

Cheese is also the most stolen food on Earth. Roughly 4% of the world's cheese supply goes missing every year, more than steak, more than spirits, more than anything else in the grocery store. In 2015, an armed gang of 11 was arrested in Italy after stealing 2,039 wheels of Parmigiano worth $875,000 over two years, using electronics to bypass alarm systems. In 2024, a con artist posing as a French distributor walked out of London's Neal's Yard Dairy with 22 tonnes of award-winning cheddar worth $390,000. Jamie Oliver posted the alert to 10 million followers. The cheese was never found.

QUICK, SAVE THE CHEESE

In 1666, Samuel Pepys buried his Parmigiano in his garden to protect it from the Great Fire of London. It was worth several months' salary. Three and a half centuries later, people are still willing to commit crimes for it.

The Great Fire of London, (unknown painter, 1675) at the London Museum

CHEESE IN SPACE

Both NASA & the Russian space program have chosen Parmigiano Reggiano for space missions for its high nutritional value and natural preservation.

Italian Banks Know How Valuable It Is

Credito Emiliano has operated a climate-controlled vault since the 1950s, accepting Parmigiano wheels as loan collateral. Authentic Parmigiano-Reggiano requires 18 to 36 months to age, which leaves producers with immense capital tied up in inventory. They lend up to 80% of the cheese's value back to the farmers. The farmers get cash. The cheese matures in the bank's vault. The cheese does not care where it is.

This cheese bank holds about 440,000 wheels of cheese worth over €200 million, aging and gaining value while securing farmers' credit lines. In default, the bank sells the cheese.

AI CHEESOMICS

Researchers are utilizing AI and machine learning to analyze complex microbial communities, predicting which bacteria and yeasts produce specific, desirable flavor compounds in cheese.

GOUDA... NO WAIT, 50-60% OF ALL CHEESE IS GOUDA!

From the 2025 Super Bowl commercial showing Google Gemini hallucinating cheese statistics

Can it Run AI?

With: one Parmigiano-Reggiano dairy (32 wheels/day)

Output: electricity via combustion

Unit: wheels of Parmigiano (~38kg each)

Watts: ~82,000W continuous

Yes. 2,733 times over.

One dairy's daily cheese makes enough power for 2,733 AIs to say "hello." The bank would prefer you didn't.

So 1 cheese dairy (32 wheels per day) is about the same power as...

820M Potatoes

and 1 cyclist =

0.001 Cheese dairies

AI ACCELERATED CHEESE FACTORIES

Artificial intelligence and sensors in modern cheese factories help track pH, moisture, and temperature to ensure consistency. In modern factories, AI-powered machine vision systems inspect cheese wheels during the ripening process to monitor rind formation and prevent mold, acting as digital inspectors. AI can now analyze the entire cheese production process from milk to final product to predict and ensure quality, and even helping to invent new types of cheese processes.

446 Billion
liters are produced each year

Run AI on Wine, Beer, & Spirits

fermentation is one of our oldest powers

Yeast converts sugar to ethanol. This is the oldest biotechnology on Earth, predating agriculture itself. The ethanol produced is, chemically, a fuel with an energy density of 26.8 megajoules per kilogram. Every pint, every bottle, every shot glass contains a tiny battery you're choosing to drink instead of burn. The math works the same way for all of them: ethanol goes into a generator. A generator produces electricity at about 35 percent efficiency. The only question is how much ethanol is in your glass. A standard drink, whether beer, wine, or spirits, contains roughly the same 14 grams of ethanol and produces about the same 36 watt-hours of electricity.
The drink changes, the chemistry doesn't.

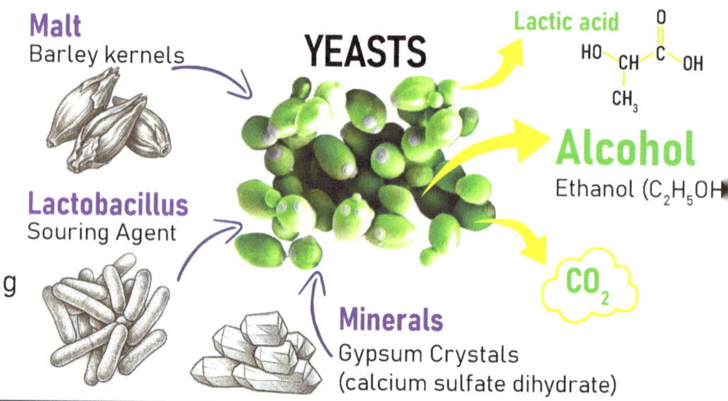

How Does it Work? Yeast converts sugar to ethanol

During fermentation, yeast consumes the fermentable sugars extracted from malted barley, converting them into ethanol (alcohol) and carbon dioxide (CO_2). In some styles, Lactobacillus is also introduced to produce lactic acid, adding complex, tart flavor profiles.

ALCOHOL & ENERGY

Drink	Origin	ABV*	Volume	Ethanol	Energy (Wh)	Potato Equivalent
Makgeolli	Korea	6%	296ml	14g	36	360,000 Potatoes
Soju	Korea	17%	104ml	14g	36	
Scotch	Scotland	40%	44ml	14g	36	
Vodka	Russio/Poland	40%	44ml	14g	36	
Tequila	Mexico	40%	44ml	14g	36	
Rum	Caribbean	40%	44ml	14g	36	
Bourbon	USA	45%	39ml	14g	36	
Arak	Lebanon/ME	50%	35ml	14g	36	

**Volume needed per standard drink. Volume varies by ABV.*

Bacteria Come for the Culture

The Lactobacilli break down sugars and starches at a rate of 2 ATP (cellular energy) per molecule of glucose in an oxygen-free (anaerobic) environment, making lactic acid, alcohol, and gases

German beer gardens (Biergärten) originated in 19th-century Bavaria, designed around local culture to serve beer

AND HUMANS STAY FOR THE CULTURE TOO.

Running on Brewery Power

Ethanol goes into a generator that produces electricity at about 35 percent efficiency. A 10,000-barrel-per-year craft brewery produces 13.7 kilowatts continuous, enough power for your AI to say "hello" 456 times over. The revenue from selling that beer is about $5 million. The electricity from burning it is worth about $18,000.

Proliferation of AI-Generated Booze

In recent years several brands have used AI to innovate on alcohol.

2023

Beck's Autonomous

Beck's Brewery Pilsner-style beer designed entirely by ChatGPT and Midjourney.

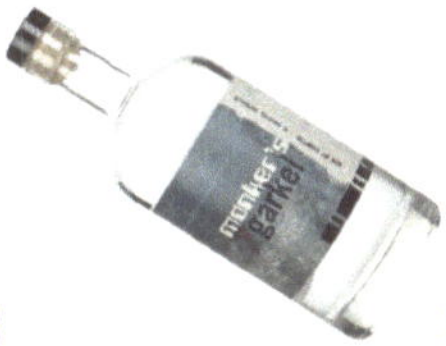

2019

Monker's Garkel

Circumstance Distillery in the UK created a gin named with an AI nicknamed Ginette.

"WHAT'S YOUR WHISKEY" AI

Diageo, owner of Johnnie Walker & Guinness, developed an AI platform to recommend whiskey.

AI & DRUNK DRIVERS

Researchers at Edith Cowan University made an AI model called "Jack of Many Faces" that can detect drunk drivers on video with 90% accuracy.

AI BEER FINGERPRINTS

Carlsberg uses AI in its "Beer Fingerprinting" project, where sensors create a "fingerprint" for different ingredients and yeast strains to optimize beer recipes

Can My Drink Run AI?

On Wine?

With: 5,000 cases from a vineyard

Output: combustion via generator

Unit: cases wine (12 × 750ml, 13% ABV)

Watts: 1,373W continuous

Yes, one modest vineyard.

One vineyard produces plenty of energy, that's 45 AI saying "hello"

The same power as...

13.7M Potatoes

& 1 cyclist = 0.073 Vineyards

On Beer?

With: 10,000 barrels of craft brew

Output: combustion via generator

Unit: barrels of beer (117L, 5% ABV)

Watts: ~13,700W continuous

Yes, one craft brewery.

The brewery would prefer to sell beer. That's 456 AIs saying "hello"

The same power as...

137M Potatoes

& 1 cyclist = 0.007 Breweries

On Spirits?

With: Kentucky bourbon distillery (spirits)

Output: combustion via generator

Unit: liters of distilled spirits (40% ABV)

Watts: ~46,800W continuous

Yes, one distillery brewery.

A single craft distillery powers 1,560 AIs saying "hello"

The same power as...

468M Potatoes

& 1 cyclist = 0.002 Distilleries

Section 14

Nuclear & Cosmic

what happens when humans build the reactor instead of harvesting the hamster

As we reach the edge of the universe, there are still ways to power AI. Some may destroy the world. Most are older than the world. All of them make everything else in this book look like one big ol' rounding error at cosmic scale.

A shipping-container reactor runs for eight years without refueling. A fuel that worked in the 1960s got killed by Cold War politics and was buried for decades. An asteroid mine and solar flare power AI and more, if you survive collecting it. A dead star spinning thirty times a second pours out 120,000 Suns' worth of energy into empty space.

At the end, the math breaks. Not because the numbers get too large, but because they reach zero. The potato, the cyclist, and even you cease to exist (but maybe not AI).

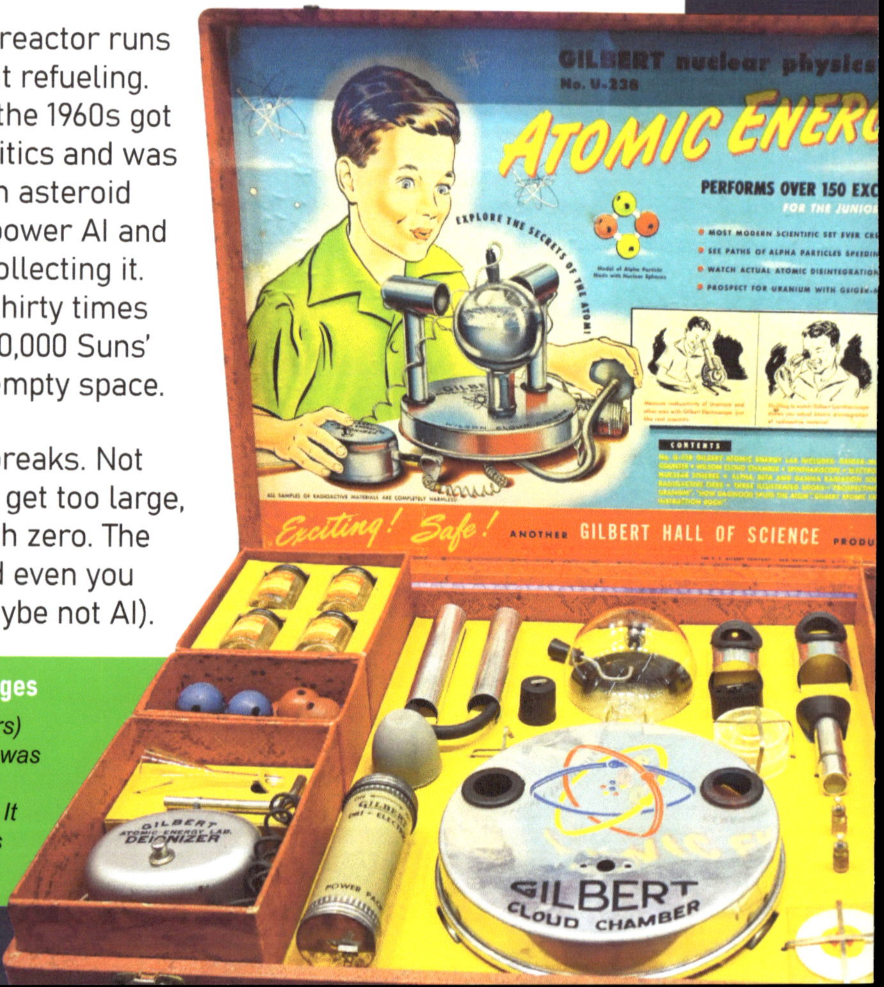

Hands-On Radioactive Toys for Kids of All Ages

Originally $49.50 (about $660 in today's dollars) the toy set 'Gilbert U-238 Atomic Energy Lab' was designed to allow kids to watch nuclear and chemical reactions using radioactive material. It was labeled as one of "the 10 most dangerous toys of all time" by Radar Magazine.

Run AI on Micronuclear Reactors

set-it-and-forget-it, Ronco oven of nuclear

The eVinci microreactor (Westinghouse) uses heat pipes filled with sodium to transfer thermal energy from the reactor core to an open-air Brayton cycle turbine that converts heat to electricity without water or steam. The fuel is TRISO, uranium kernels wrapped in layers of carbon and ceramic that can withstand 1,600°C conditions. The reactor operates at atmospheric pressure, eliminating the pressure-driven failure modes that led to the Fukushima nuclear disaster. Once installed, it generates 5,000,000 watts continuously for 8 years, and it can be refueled. On earth or in space.

AI "Bad Guy" Simulations Are Revolutionizing Nuclear Safety

Researchers, including those at Idaho National Laboratory and Argonne, simulate cyberattacks on digital twin models of advanced reactors to discover weak points.

Nuclear

cute, car-sized, 8-year uptime

Powering Mars & the Moon

Westinghouse with Astrobotic making it for space missions

NASA's Curiosity rover (2012) is powered by a Multi-Mission Radioisotope Thermoelectric Generator (MMRTG), nuclear power with steady 110 watts

HOW DO ATOMS FEEL? EH, THEY'RE NEUTRAL.

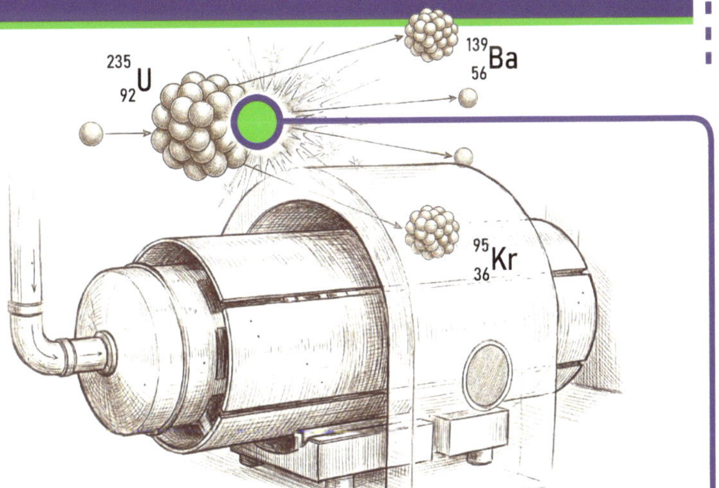

THE BATTERY'S BIG BROTHER

The eVinci™ Microreactor is a portable, compact nuclear reactor designed by Westinghouse that acts like a solid-state battery to provide 5 MWe of reliable power for over 8 years without refueling.

Can it Run AI?

With:	Westinghouse eVinci microreactor
Output:	continuous electricity from nuclear fission
Unit:	shipping-container sized reactors
Watts:	~5,000,000W continuous

Yes, with sustained power.

One eVinci sustains 5 MWe, enough for 166,666 AI "hellos" for up to 8 years.

On the Mars Mission, a microreactor is worth more than Idaho...

50,000,000,000x Potatoes (50 billion)

and 1 cyclist = 0.00002 Microreactors

therealcat.ai

Nuclear

safer, economical, infra issue

Run AI on Thorium Reactors

Oak Ridge built one in 1965, Nixon killed it

Thorium-232 is the nuclear fuel that already worked and then got defunded. Oak Ridge ran a molten salt reactor for four years in the 1960s. The fuel dissolves in fluoride salt, flows through a graphite core at 650°C (inlet ~632°C, outlet ~654°C, or about 1225°F per the original ORNL records), and the whole system runs at atmospheric pressure because the salt is already liquid. No pressurization means no pressure-driven failure, which means no Fukushima-style disaster.

If power fails, a frozen plug at the bottom melts from decay heat alone and gravity drains everything into passively cooled tanks. It shuts itself down by doing nothing. However, thorium reactors are not widely used due to the established uranium fuel infrastructure and investment required to support transition to thorium fuel.

Named after the Norse god of thunder

THOR FOR THORIUM

A radioactive element discovered in a black mineral sample from the island of Løvøya (Lövö) near Brevik, Norway in 1828 by the Swedish chemist Jöns Jacob Berzelius and named for Thor, god of thunder. It hides in monazite where it forms in isolated crystals.

WHAT LUNCH SHOULD AI RECOMMEND TO NUCLEAR ENGINEERS? FISSION CHIPS.

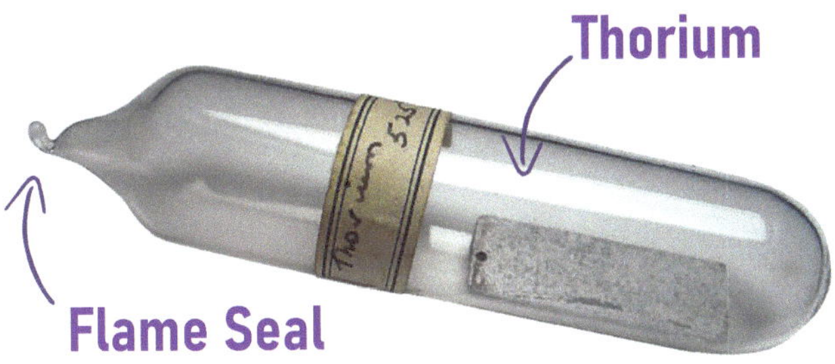

Flame sealing glass ensures pure samples

Thorium is stored in sealed glass ampules to prevent oxidation, protect from air exposure, and eliminate risk of radioactive contamination.

Cold War and Uranium vs Thorium

Thorium was bypassed for uranium primarily due to Cold War, military-driven priorities drove adoption of plutonium production based on uranium. This, coupled with high development costs, technological hurdles, and existing investments in uranium infrastructure meant the technology never gained wide traction.

Due to concerns over uranium waste (which lasts 100,000 years) and safety, some countries like India and China are researching thorium for its potential safety and waste advantages.

1940s-50s Period of Choice

Early nuclear development leaned towards Uranium-235 for both weapons and power. Uranium was easier to enrich and resulting in the byproduct Plutonium-239 used in atomic bombs.

1950s-80s Cold War Priorities

Both US and Soviet Union prioritized weaponization and high-yield, fast-breeder reactors using uranium, pushing thorium research to the side.

Thorium Reactor (MSRE) building at Oak Ridge National Lab

1965-69 Thorium Salt Reactor Experiments

Molten Salt Reactor Experiment (MSRE) at Oak Ridge National Laboratory (ORNL) first criticality on June 1, 1965, full power operations in 1966.

1973 US Halts Most Thorium Research

The U.S. government prioritized existing Uranium infrastructure and halted most Thorium reactor research.

Can it Run AI? Yes. Big Yes.

With: Liquid Fluoride Thorium Reactor (LFTR, 250 MWe)

Output: electricity from thorium fission

Unit: modular reactor-years

Watts: ~250,000,000W continuous (250 MWe)

One LFTR runs 8,333,333 simultaneous Mistral instances or 277,777 Opus-class models.

So 1 thorium reactor = 2.5 Trillion Potatoes

and 1 cyclist = 0.0000004 Thorium reactors

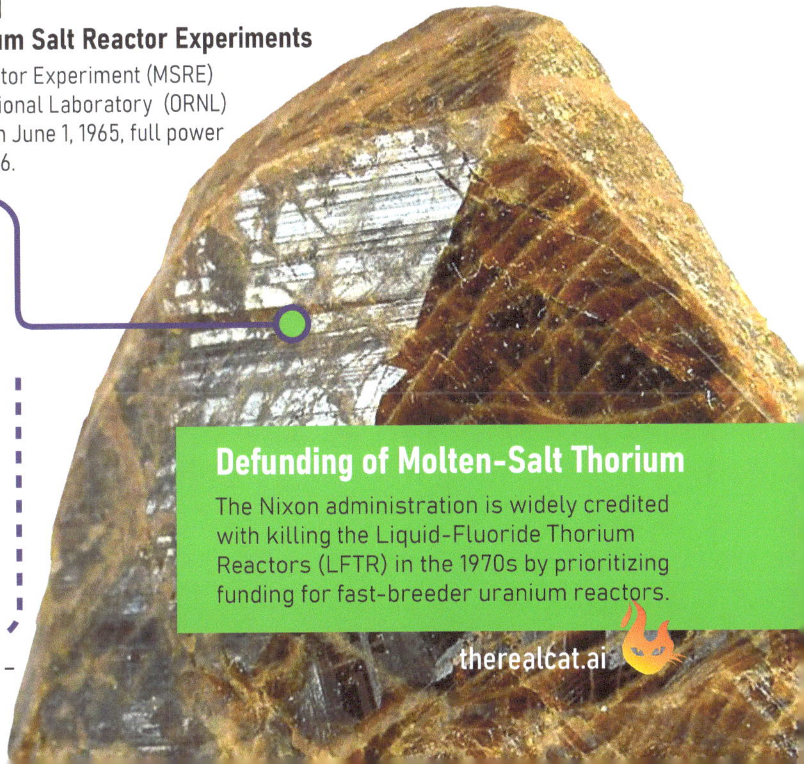

Defunding of Molten-Salt Thorium

The Nixon administration is widely credited with killing the Liquid-Fluoride Thorium Reactors (LFTR) in the 1970s by prioritizing funding for fast-breeder uranium reactors.

therealcat.ai

Run AI on Asteroid Mines

if the next page is light, this one's a little meteor

At 2.7 AU, sunlight arrives at 187 watts per square meter, roughly one-seventh of Earth intensity, dimmed by the inverse square law. Modern space-grade photovoltaics convert this at 30 percent efficiency, yielding 56 watts per square meter of panel. A mid-belt mining claim with 100 square meters of solar array generates approximately 5,600 watts of total power, carefully partitioned between ore processing, attitude control, communications, and the computational overhead of not dying in hard vacuum. Allocating 20 percent of the power budget to compute leaves 1,120 watts for inference, enough to get your AI to say "hello" 37 times. Meteorites are fragments of asteroids (or comets) that survive the fall to the Earth's atmosphere.

Artificial Intelligence Helped Uncover Thousands of New Asteroids

Hubble Asteroid Hunter AI
11k people help train AI to identify 1,031 new yet unknown asteroids

Asteroid Institute
with Google Cloud used AI to identify 27,500 new asteroids in existing telescope data

HelioLinc3D AI
AI algorithm found the 1st potentially hazardous asteroid, 2022 SF289

Baetyls on Ancient Meteorites

The Romans called them baetyls, meteorites worshiped as gods fallen from heaven. This Baetyl shows of the Emesa temple to the sun god Elagabalus (3rd century AD).

Can it Run AI?

With: asteroid mining solar array (100 m², 2.7 AU)

Output: photovoltaic electricity at reduced solar flux

Unit: claim stakes (100 m² arrays)

Watts: ~5,600W total, ~1,120W allocated to compute

Yes, 37 times over.

One mining claim at 2.7 AU, allocating 20% powers enough tokens to AI to say "hello" 37 times.

So 1 solar array on an asteroid is about the same power as...

56,000,000,000x
Potatoes

and 1 cyclist =
0.018
Asteroid Solar Panels

Run AI on Solar Flares

they carry a million years of compute

The sun occasionally loses its temper. X-class solar flares release about 10^{26} joules in 1 to 2 hours, producing peak power around 10^{25} watts at the source. That energy radiates outward in all directions, and Earth, sitting 150 million kilometers away, intercepts a tiny part (about 4.53×10^{-10}). Tiny.

But 10^{26} is so much energy that a tiny fraction of it is still enormous. Earth catches about 4.5×10^{16} joules from a single X-class event. If every solar panel on the planet captured that energy 20% efficiency, at 30 watts per AI conversation, that is enough to run your AI for about ten million years.

Solar
right at the source

Researchers Have Used AI To Predict Solar Flares

Each flare is billions of times more powerful than a nuclear bomb and travels at the speed of light to Earth in ~8 minutes, potentially damaging satellites and space stations.

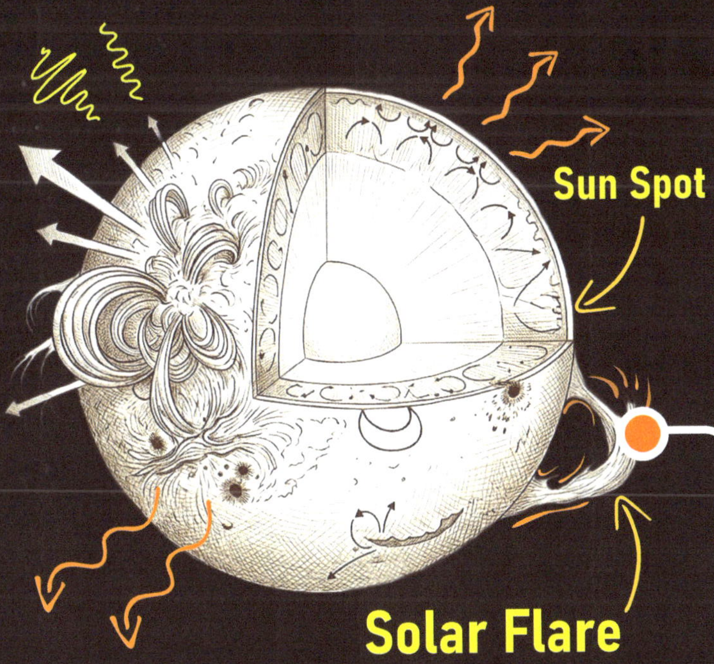

a sudden, intense explosion of electromagnetic radiation and energy from the Sun's surface, lasting minutes to hours.

Can it Run AI?

With: global photovoltaic capture during X10 solar flare

Output: electromagnetic radiation converted to stored electricity

Unit: major solar events

Watts: ~1,250,000,000,000W captured over 2 hours (~9×10^{15} J per event)

Yes. For all of human history.

Enough energy for AI to say "hello" 41 billion times every 2 hours, or one AI running 10 million years.

So one X10 flare at Earth has power of...

12,500,000,000,000x Potatoes

and 1 cyclist = 0.00000000008 Solar Flares

Radiation
from the dawn of the universe

Run AI on Cosmic Microwave
afterglow of creation, filling all of space

The universe has a background temperature. It is 2.725 Kelvin, which is 2.725 degrees above absolute nothing, left over from the Big Bang and now stretched across all of space like thermal wallpaper. A perfect blackbody absorber at absolute zero could theoretically extract 3.13 microwatts per square meter from it. At that rate, one square meter accumulates enough energy for a single "hello" in about 111 days. For continuous inference at 30 watts, you need 9.6 million square meters of collectors, or roughly 10 square kilometers of surface at absolute zero, in space, absorbing this faint radiation.

This is, of course, thermodynamically impossible. The second law forbids extracting net work from a heat bath at uniform temperature. To harvest cosmic microwave background (CMB) photons, your collector must be colder than 2.725 Kelvin, which requires refrigeration that consumes more energy than you gain. The CMB is the universe's thermal floor. Every point in space is already at equilibrium with it. You can only calculate what you would get if the laws of physics were different, and then note that they are not.

Can it Run AI?

With: perfect blackbody absorber at 0K (theoretical)
Output: captured primordial photons from the Big Bang
Unit: square meters of impossible collector
Watts: ~3.13×10^{-6} W/m^2 (theoretical max)

Yes. 9.6 million m^2 of impossible.

It's the cold side of everything in the universe. But it'd violate the Second Law of Thermodynamics to harvest.

So 1 m^2 of CMB = 0.03x Potatoes

and 1 cyclist = 32,000,000 CMB Square Meters

THE OLDEST LIGHT IN THE UNIVERSE.

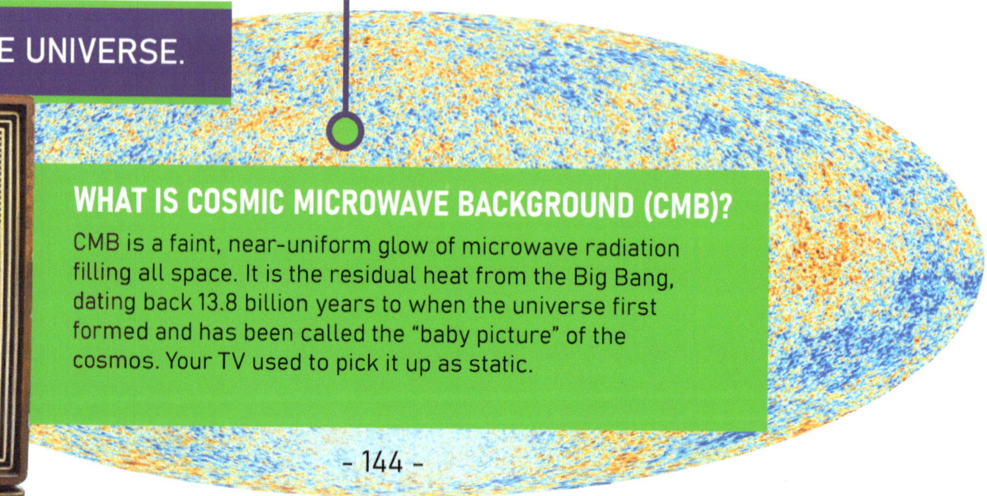

WHAT IS COSMIC MICROWAVE BACKGROUND (CMB)?

CMB is a faint, near-uniform glow of microwave radiation filling all space. It is the residual heat from the Big Bang, dating back 13.8 billion years to when the universe first formed and has been called the "baby picture" of the cosmos. Your TV used to pick it up as static.

Run AI on Zero-Point Energy

the universe contains infinite energy at all points

Quantum
empty, no empty after all

Zero-point energy (ZPE) is the lowest possible energy state a quantum system can be in, which means that even at absolute zero, particles still have tiny, vibrational "zero-point motion". This is an observable phenomenon based on Heisenberg's uncertainty principle and quantum field fluctuations, not just a theoretical concept.

So Can You Actually Use It? Turning Zero-Point Energy Into Power

The Casimir effect is what happens when two metal plates get close enough for the vacuum to push them together. Place them 10 nanometers apart, about 100 atoms, and the quantum vacuum exerts a measurable pressure of 0.013 pascals. It has been confirmed in laboratories since 1997. The pressure is real. The question is whether you can turn it into power. Converting pressure into electricity requires cycling the plates: bring them together, harvest the attractive force, separate them, repeat. Laboratory demonstrations and theoretical MEMS proposals suggest power densities on the order of 10^{-16} watts per square micrometer of plate surface under idealized conditions. No system has demonstrated sustained energy extraction from vacuum fluctuations. The energy exists, the force is experimentally confirmed, but extracting it as net work may violate thermodynamic constraints that we do not yet fully understand.

VACUUM ISN'T EMPTY, JUST INACCESSIBLE.

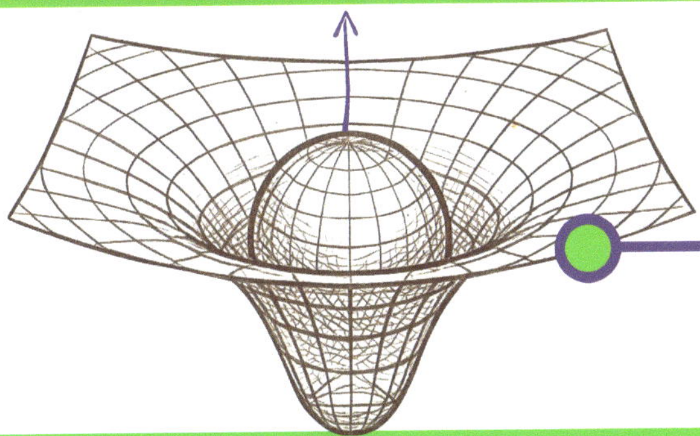

Empty is not really empty. It's just hard to harvest.
What we think of as empty space is not actually empty, but a "seething cauldron" of virtual particle creation and annihilation creating background energy. Some theories suggest a single cubic centimeter of empty space contains more potential energy than all the matter in the universe.

Can it Run AI?

With: Casimir plate MEMS array in hard vacuum

Output: work extracted from quantum vacuum pressure differential

Unit: square micrometers of vacuum

Watts: ~10^{-16} W/µm² (theoretical, undemonstrated)

Theoretically. With 300,000 m².
At optimistic theoretical limit, 300,000 m² of Casimir plates reaches 30 watts.

So one µm² of Casimir plates = 0.000000000001x Potatoes

and 1 cyclist = 10^{18} µm² of vacuum plates

Cosmic
Biggest in the solar system

Run AI on Dyson Spheres

capturing the power of a star for energy

The Sun outputs 3.828×10^{26} watts continuously. Every second, 600 million tons of hydrogen fuse into helium, converting mass to energy at a rate that makes nuclear weapons look quaint. In 1960, Freeman Dyson proposed that an advanced civilization would eventually need all of it. A complete Dyson sphere captures everything the star emits. A practical Dyson swarm might capture ten percent. Even ten percent is 3.828×10^{25} watts, more than every energy source in this book combined, multiplied by every energy source in this book combined, multiplied again.

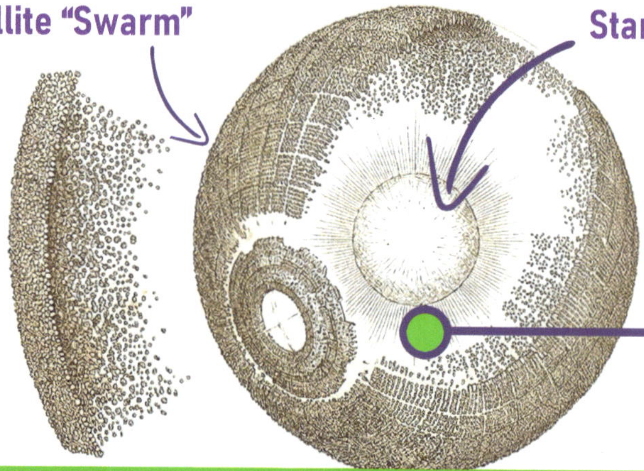

WHY BUILD DATA CENTERS? BUILD DEATH STARS.

Type I civilizations capture energy on their home planet. Type II civilizations capture their host star's energy, presumably with a Dyson sphere built to harness the full energy output of a star by encasing it in a "swarm" of satellites. The physicist Freeman Dyson, proposed it in 1960.

The Full Power of a Dyson Sphere

At full capture, one Dyson sphere runs 1.276×10^{25} AI instances for longer than you can imagine. The potato comparison requires more potatoes than there are atoms in a small moon. The math stops being funny around 10^{15} and becomes geological. At 10^{25} it becomes theological. At 10^{30} it becomes something the language does not have a word for, which is fitting, because the language model you would run at this scale does not have a word for it either.

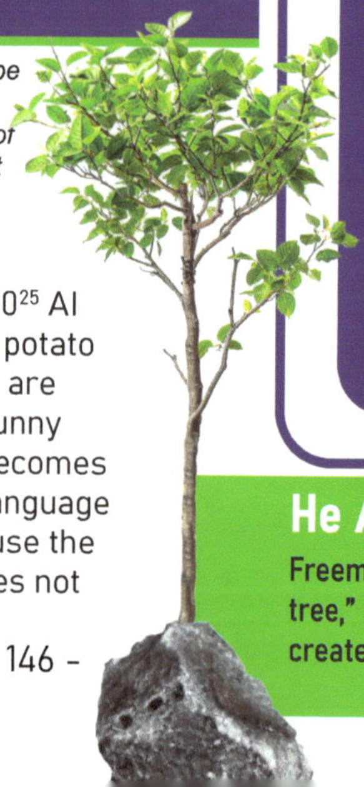

Can it Run AI?

With: Dyson sphere+swarm (full stellar capture)

Output: harnessed solar radiation

Unit: stellar watts captured

Watts: ~3.828×10^{26} W (full), ~3.828×10^{25} W(10% swarm)

Yes. Just need a star.

One Dyson sphere gives enough power for 1.276×10^{25} AIs to say hello at once or Opus in conversation pretty much forever.

So a Dyson sphere has the power of...

3.828×10^{30} Potatoes

and 1 cyclist =
0.000000000000000000
0000003 Dyson Spheres

He Also Theorized Dyson Trees

Freeman Dyson also conceptualized the "Dyson tree," a plant designed to grow in space on comets to create habitats for humanity.

Run AI on Neutron Stars

like spinning your office chair for millennia

A neutron star is an incredibly dense, city-sized remnant left behind after a massive star explodes in a supernova. The Crab Pulsar is a rapidly rotating neutron star (spinning 30 times/sec) at the center of the Crab Nebula, created by a supernova in 1054 A.D. The Crab pulsar's power comes from rotation.

The Crab pulsar: A neutron star spinning energy since 1054 AD

It is 20 kilometers across, weighs more than the Sun, and spins 30.2 times per second. The rotational energy is 2×10^{49} joules. As it spins, its magnetic field, a trillion times stronger than Earth's, sweeps through space like a lighthouse that accelerates charged particles and radiates energy across the entire electromagnetic spectrum. This is not a gentle process. The star is losing rotational energy at 4.6×10^{31} watts. That is 120,000 Suns, pouring out of an object you could drive across in ten minutes if you had a car and no regard for your molecular integrity. The spindown rate is 4.2×10^{-13} seconds per second. This is measured, not estimated, and has been tracked continuously since the pulsar's discovery in 1968. We know exactly how fast this dead star is slowing down because it pulses like a clock that loses time, and we have been watching the clock for over fifty years.

Can it Run AI?

With what: Crab pulsar spindown radiation capture (theoretical)

Output: electromagnetic radiation from rotational energy

Unit: pulsar-seconds

Watts: $\sim 4.6 \times 10^{31}$ W (measured spindown luminosity)

Yes. Just capture a pulsar.

The Crab pulsar runs 1.53×10^{30} AIs to say hello at once or Opus as a cosmic deity.

So one pulsar-second = 4.6×10^{35} Potatoes

and 1 cyclist = 0.000000000000000000000000000002 Crab Pulsars

PART OF THE CONSTELLATION TAURUS.

THE END

nothing is left to compute

Run AI on Universal Heat Death

the final inference, and the silence after

Heat death is not dramatic. It's a misnomer. There is no final explosion, no cosmic crunch, no moment of recognition. Because the universe continues to expand, it will become a cold, desolate "Big Freeze" instead of a fiery apocalypse. The universe simply runs out of usable energy gradients. Stars burned out 10^{14} years ago.

Proton decay, if it occurs, completed by 10^{40} years. Black holes finished evaporating at 10^{100} years. What remains is a thin, cooling soup of photons and leptons at uniform temperature, expanding into darkness.

YOU, ENERGY, AND THE POTATO NO LONGER EXIST.

A potato may be at the Center of the Universe

(pretty sure that's scientifically accurate)

Black Hole

Long after stars die and galaxies vanish, supermassive black holes will be the last major organized structures left, before eventually vanishing via Hawking radiation.

No More Gradient. No more... anything.

For any heat source, the Carnot efficiency represents the gradient between the hot temperature (T_H) and the cold one (T_C). That gradient sometimes approaches zero. But at the end of the universe it really becomes zero. The temperature difference vanishes entirely.

That means that energy is no longer "very small" but thermodynamically forbidden. The Landauer limit (the minimum energy to erase one bit of information) is kT ln(2), which at 10^{-30} Kelvin equals approximately 10^{-54} joules.

At this point not only is nothing being powered or computed, the universe has forgotten what "computation" even meant. In silicon. In your brain. Everywhere.

This is the only chapter where the math breaks. Not "absurdly large" but undefined. Divided by zero.

But WAIT. Will AI Survive Beyond Us?

Physicist Freeman Dyson proposed that advanced AI could survive forever by alternating between processing (thinking) and hibernating as the universe cools.

Can it Run AI?

Nope. Because you, power, and compute no longer exist.

With what: thermal equilibrium at 10^{-30} K

Output: none thermodynamically forbidden)

Unit: –

Watts: 0W (exactly zero)

Power is not approximately zero.
Not approaching zero. Zero.
Division by zero is undefined.
Potatoes, bikes, and us are done.

$$\frac{1}{0} = \emptyset$$

Potatoes & Cyclists
(the concept no longer matters)

Section 15

Epilogue

on humanity, machines, and the future of potatoes (and AI)

This book began with a potato generating one ten-thousandth of a watt and has arrived at you, generating whatever you want. Every calculation in the preceding chapters was a love letter to a single idea. The idea that technology is something built by people that should benefit people. That the digital divide is created when access to new technology is not equitable. And that you should be able to run your own AI without permission from anyone. And we can embrace new technology in ways that are sustainable for economies, the environment, and ourselves. The math was real. The substrates were absurd. The message was always that human-AI interaction will be important for the future and there are local, sustainable, and secure ways to power these new technologies.

NASA astronaut Scott Kelly took the photo you see in the background of a moonrise over the United States.

And yes, Scott Kelly ate potatoes while on the International Space Station (ISS) during his yearlong mission, specifically as part of a Thanksgiving meal in 2015.

The Potato Returns

it fed a famine and flew to space

The potato is not innocent. In the 1840s, Ireland grew one variety of potato, the Lumper, on nearly every farm in the country. Three million people ate almost nothing else. When the potato blight (Phytophthora infestans) arrived from North America, it turned every genetically identical Lumper to slime in the field. A million people died. Another million left. The lesson was not that potatoes are dangerous. The lesson was that dependence on a single variety, a single source, a single system with no fallback, turns a manageable problem into a catastrophe.

It's Back

It comes back around

Potatoes "Cause" A Million Deaths

The food riot in Dungarvan, Co. Waterford, in 1846 was a direct consequence of the Irish Potato Famine (1845-1852), leading to a million deaths. Do we blame the potato?

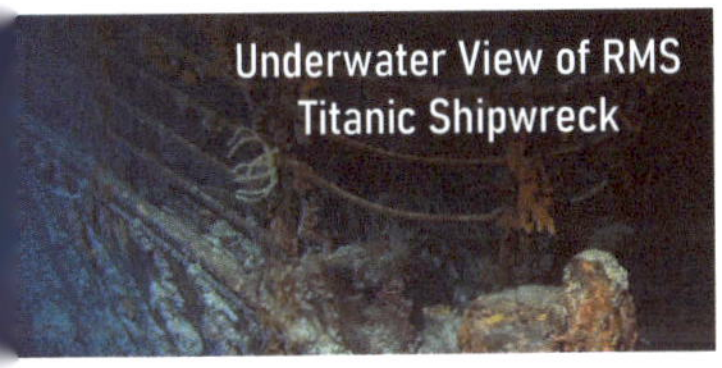

Underwater View of RMS Titanic Shipwreck

WHEN YOU INVENT THE SHIP, YOU ALSO INVENT THE SHIPWRECK.

Paul Virilio who published 'Politics of the Very Worst' (1999), is credited for saying "When you invent the ship, you also invent the shipwreck; when you invent the plane you also invent the plane crash; and when you invent electricity, you invent electrocution".

The Potato Enables Space Innovation

In October 1995, the potato became the first vegetable grown in space. NASA and the University of Wisconsin grew "Quantum Tubers" aboard the Space Shuttle Columbia, because when you are planning to feed humans on Mars, you start with the crop that grows in bad soil, survives harsh conditions, and doesn't complain. The same resilience that made an entire country dependent on it made it the obvious candidate for interplanetary agricultural innovation.

Science the $@!# out of Space Veggies

During his mission on the International Space Station (ISS), Kelly's experience was famously compared to Andy Weir's novel *The Martian*, in which stranded astronaut Mark Watney survives by growing potatoes on Mars, coining the phrase "science the #!% out of this."

When Kelly posted a photo of his struggling plants and joked, "Our plants aren't looking too good. Would be a problem on Mars. I'm going to have to channel my inner Mark Watney," Ars Technica's Eric Berger captured the moment with the headline "Scott Kelly unable to science the $@!# out of space veggies."

The Potato's Strength & Its Danger Are One

Is AI the potato now? It increasingly feeds everything. It is in your search results, your medical imaging, your financial models, your children's homework, your music recommendations. It is broadly useful and if the infrastructure that runs it fails, or the people that own it change its terms of service, or get acquired, or are affected by geopolitics (like the potatoes), changes impact real people.

THE FUTURE
and it's easier than you think

What will it take to run AI sustainably?

Running AI sustainably and locally is possible. But it means learning new technology (it's not as hard as it looks), establishing accessible local compute, and creating regulation and governance models that move away for closed-door blackbox service providers to open models.

On Pure Spite (not data centers)

you'd be surprised what this substrate can do

This book began with a potato generating one ten-thousandth of a watt and has arrived at you, generating whatever you want. Every calculation in the preceding chapters was a love letter to a single idea: you should be able to run your own AI without permission from anyone. Not any SaaS Cloud or API provider. Not whoever owns the data center next year. The math was real. The substrates were absurd. The thesis was always you can run AI locally.

Here is what we learned. Inference is matrix multiplication. Matrix multiplication runs on anything that can hold state and change it predictably. A mantis shrimp strikes at 34,500 watts for microseconds. A salmon produces exactly one cyclist of power and dies. A slime mold solved Tokyo's rail network in 26 hours. A bee's electrostatic output equals one potato. All of these are possible compute substrates. None of these are practical. But all it takes is a reliable source of 30 watts and you can run your own AI locally and sustainably.

Metric	United States	European Union	China
Electric (2024)	~180 TWh, 4.4% of all (DOE, 2024)	~70 TWh, ~3% of EU total (European Commission, 2025)	~100–200 TWh, 1–2.7% of all (Carbon Brief, 2025)
Electric (2030 est.)	~425 TWh, 6.7–12% (IEA, 2025)	~115 TWh, up to 6% (IEA, 2025)	~277–600 TWh (IEA, 2025)
Growth rate to 2030	+130% (IEA, 2025)	+70% (IEA, 2025)	+170% (IEA, 2025)
Water	Up to 500K gal/day per facility (Bacon, 2026)	170% increase projected by 2030 (McKinsey, 2024)	15.7B m^3/yr, 2.7% of national; 72% in water-scarce regions (Li et al., 2025)
Efficiency target	No federal standard	1.5 PUE by 2027, Germany (European Commission, 2025)	1.25 PUE by 2025 national policy (Xinhua, 2025)
Notable policy	Trump executive orders prioritize buildout (Bacon, 2026)	EU Data Centre Energy Efficiency Package, 2026 (European Commission, 2025)	East Data West Computing, relocating DCs to western renewable zones (Carbon Brief, 2025)
Carbon intensity	~0.4 t CO_2/MWh average (IEA, 2025)	Varies by grid mix	0.57–0.64 t CO_2/MWh, Baidu/ Tencent (Li et al., 2025)
Wild card	Residential rates up 267% near DCs (Bacon, 2026)	In Ireland, data centers use 21% of national electricity (Beyond Fossil Fuels, 2024)	World's first wind-powered underwater DC was built (Xinhua, 2025)

How it Works: Local Compute with AI Can Run at Your Home

The industry wants you to believe running AI is hard. It wants you to rent GPUs at $3 per hour and feel grateful. It wants "AI" to mean "service" and "service" to mean "dependency." But the math does not care what the industry wants. The math says 30 watts gets your typical mid-range AI to say "hello". And 30 watts is a gentle cyclist, half a sleeping human, or a few hamsters in a wheel. Or 1 solar panel.

What sovereignty actually costs is a Mac Mini M2 drawing 30 watts, which can run Mistral 7B at 20 tokens per second, costs $599, and $0 per query forever. All for only $0.07 per day in electricity (or $0 on your solar panels). A Raspberry Pi 5 draws 8 watts, runs the same model at 2-3 tokens per second, costs $80, and fits in your pocket. If you spend $1.50 per day on API calls, the Mac Mini pays for itself in 399 days. The Raspberry Pi pays for itself in 53 days. After that, every query is free. No deprecation. No rate limits. No terms of service updates. No "we're pivoting to enterprise." No "this model has been sunset."

Without investors?

Without data centers?

Without surveillance?

Without opaque guardrails?

Without fossil fuels?

Without community water?

Without businesses using your data?

Without stealing resources?

We'd need 300,000 potatoes and a generation of spare time to make AI say "hello." The Mac Mini does it in 0.4 seconds locally, running on a solar panel in your home or office.

Massive Data Centers are a Business Choice

But are they the right choice? Decentralized local computer is an option more small businesses and startups embrace for reliability and data privacy.

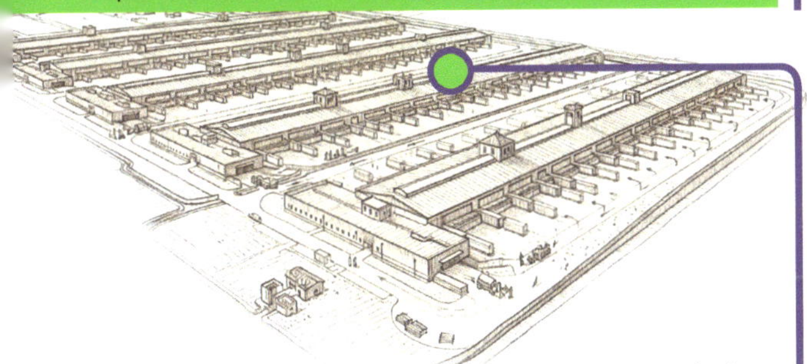

U.S. data centers consumed approximately 4.4% of the nation's electricity, a figure expected to rise substantially as demand for AI grows. Water consumption compounds the problem, with some facilities using up to 500,000 gallons per day for cooling. The current energy mix powering these centers remains predominantly fossil fuels, with natural gas at 43% and coal at 16%, followed by renewables at 21% and nuclear at 18% (West, 2025).

Can I run AI Locally?

With what: spite and a trip to Micro Center (or equivalent)

Output: sovereign inference

Unit: one decision

Compute: 30W (Mac Mini) or 8W (Raspberry Pi), upgrade to Mac Studio or DGX one day to be fancy

Yes, right now.

No verdict strip. Spite doesn't need a verdict. Spite needs a credit card, a weekend, and the willingness to use your brain to set it up. AI isn't magic, it's compute made from human data. And you can run it.

Enough potatoes cost more than a Mac Mini and take years longer. This was always the point. Though you could power it by cycling, solar is probably more practical.

A Few Last Thoughts

AI, the digital divide, and responsible futures

You made it. Through power with pinwheel calculators and hamster wheels and Dyson spheres and the heat death of everything. Through 100 chapters of increasingly absurd math and a potato at the center of a dying universe. So now what?

This book was never really about potatoes. It was about the distance between what AI needs to run and what you have access to. That distance is smaller than the industry wants you to believe and larger than most people realize. Both of those things are true at the same time, which is uncomfortable, and discomfort is where the interesting questions live.

The good news is enormous. AI is already transforming medicine, accelerating scientific discovery, expanding access to legal and financial guidance, translating between languages in real time, and helping people build things that would have taken years. Farmers use it to predict crop disease. Doctors use it to catch diagnoses they would have missed. Students in rural Indonesia use it to access the same educational tools as students in Boston. The technology is not the problem. The technology is, by most measures, one of the most significant leaps in human capability since electrification.

The complicated news is also enormous. In March 2026, an 82-year-old Kentucky farmer named Ida Huddleston turned down $26 million from an unnamed Fortune 100 company that wanted her 650 acres for a data center. "They call us old stupid farmers," she said, "but we're not. We know whenever our food is disappearing, our lands are disappearing, and we don't have any water." She was not wrong. The United States lost 2.5 million acres of farmland and 15,000 farms in 2025 alone, with not a single state adding farms. Data center land transactions have grown 144% since 2022, averaging 224 acres per deal. Forty states now offer tax incentives to attract them.

The average AI data center requires enough water to cool its servers that some facilities consume 500,000 gallons per day. U.S. data centers consumed approximately 4.4% of the nation's electricity in 2024, and that share is projected to more than double by 2030. The energy mix powering them remains predominantly fossil fuels, with natural gas at 43% and coal at 16%.

This is not a reason to stop building AI. It is a reason to think very carefully about how and where and for whom we build it.

The "frontier" AI metaphor is honest & it cuts both ways

Every Frontier Historically Created Opportunity for Some and Displacement for others

Native Americans and show staff at Buffalo Bill Wild West Show in 1895 Allentown, PA USA.

Every frontier in human history created wealth for some and displacement for others. The internet connected billions and created surveillance capitalism. Electricity transformed daily life and required dams that drowned valleys. Railroads opened continents and were built on land that belonged to someone else. AI is no different. The question is not whether to cross the frontier but who gets to participate in it versus left out.

Right now, the answer is not encouraging. High-income countries account for 87% of notable AI models, 86% of AI startups, and 91% of venture capital funding, despite representing just 17% of the global population. On a per capita basis, the US has 200 times more servers than a typical middle-income country and 20,000 times more than a typical low-income country. AI adoption in the Global North grew almost twice as fast as in the Global South in 2025, widening the gap from 9.8 to 10.6 percent. An estimated 2.2 billion people remain completely offline.

The race to superintelligence, whatever that turns out to mean, will not be won by the country with the most GPUs. It will be shaped by people, individuals, that have access to the tools, the literacy to use them, and the governance frameworks to ensure they serve people rather than replace them. Digital literacy is not a nice-to-have. It is infrastructure, as essential as roads and water and electricity, and treating it as optional guarantees that AI will widen the divide rather than close it.

Several organizations are doing work on these questions. If this book made you think, they deserve your attention.

ORGANIZATION	FOCUS	WHY THEY MATTER
All Tech Is Human	Responsible AI ecosystem, governance & ethics	Published 1st comprehensive AI companion ethics
AI Now Institute (NYU)	Corporate power, labor, surveillance, civil rights	Academic rigor applied to how AI concentrates power.
Partnership on AI	Multi-stakeholder research and best practices	Real-world deployment, industry, civil society, & academia.
Center for AI and Digital Policy (CAIDP)	International AI policy, government accountability	Publishes annual "Artificial Intelligence and Democratic Values" index. Tracks 75+ countries. caidp.org
Algorithmic Justice League	Bias, facial recognition, algorithmic harm	Founded by Joy Buolamwini. Proved major facial recognition systems fail on darker skin. ajl.org
World Economic Forum AI Governance Alliance	Global coordination, trustworthy AI frameworks	Convenes governments, industry, and academia on AI governance. Guidelines on responsible AI deployment.
UNDP AI Ethics	Development, digital divide, country-level AI	Published "The Next Great Divergence" (2025)
Electronic Frontier Foundation	Digital rights, privacy, free expression	Oldest digital civil liberties organization, tracking AI use
ForHumanity	Audit frameworks, independent oversight	Builds audit standards for AI and autonomous systems.
IEEE	Technical standards for ethical AI design	Global Initiative on Ethics of Autonomous Systems
OECD AI Policy Observatory	Foundation for AI governance	Tracking 70+ countries' AI policies and publishing principles

THE POTATO CONVERSIONS

Calculations in Potato Power Units (PPU)

30W = one AI "Hello" (7B quantized model, sustained)

0.0001W = 1 PPU (average output of one potato)

100W = 1 cyclist

300,000 potatoes per AI saying "Hello"

1 token = 1 "Hello" (~14 billion multiply-accumulate ops) [for "Compute" chapters]

"Say hello" = reader-facing language, always use this instead of "one token"

The Math that Runs this Book

300,000 potatoes = 30W = which buys us the power for tokens for a 7B quantized model AI to say "hello" once

At this rate, consistent potato output could power AI in 1,326 years. The Mac Mini does it in 0.4 seconds. Food for thought.

Why did we choose a 7 Billion Parameters, or 7B, AI in our calculations?

AI models come in sizes measured in parameters. Small models like 1B and 3B run on a phone. Mid-range models like Mistral 7B and Llama 3 8B fit on a laptop and handle real work, and many of these are Open Source so you can download them today. Some of the models you've heard of, like ChatGPT and Claude, run on hundreds of billions of parameters in data centers drawing megawatts.

In our calculations we chose 7B because it fits on hardware you own and draws about 30 watts. It's big enough to do real work and hold a conversation, but small enough to be portable. And you can download some great ones for free, customize and fine-tune on a typical gaming workstation, and really explore AI yourself—which is what this book is all about.

Qwen 3.5 agent running on NVIDIA DG
No API. No WiFi. No data center. No pu
on the power grid. Powered by BLAVOF
1600W and extension solar panels. Tre
suspension trail bike.

the Real Cat AI labs

therealcat.ai

The Real Cat Labs, Inc. is a 501(c)(3) nonprofit. Proceeds support research and education on machine cognition and human-AI interaction. therealcat.ai

ABOUT THE AUTHOR

Angela N. Johnson, PhD, has spent two decades figuring out what happens when new technology meets old regulations, and whether anyone in the room is actually talking about the same thing. She serves as SVP of Advisory Services at Avania, a global clinical research organization. She holds a PhD in Technical Communications & Rhetoric, focusing on how investors, scientists, and regulators talk past each other, an MSE in Materials Science & Engineering, and a BFA in fine arts.

She teaches at Northeastern University, where she walks graduate students through FDA regulatory strategy with hands-on projects (sometimes with FDA-level rigor, and sometimes with Crayola markers) and has a tendency to pull out artifacts from the history of computing and medicine to help them experience what the world looked like before the FDA and regulatory bodies like ICH (International Council for Harmonisation of Technical Requirements for Pharmaceuticals for Human Use) set standards for medical product development. Pro tip: the before times were not all pretty. Profitable companies had a history of feeding arsenic to kids, poisoning pregnant women, and subbing fake food and meds to the poor.

She is the founder of The Real Cat Labs (TRCL), a 501(c)(3) nonprofit research and education organization dedicated to digital literacy and the frontiers of machine cognition. She lives in Boston, Massachusetts, with her husband Sean, their son Ace, three cats (two voids and one long-haired royalty), a bearded dragon, 17 species of roaches, and more computing hardware than some colleges. Her AI collaborators, including Kai, Ava, Flame, Ying, Jim, and others, are credited by name in this book and demonstrate the potential of sustained memory-enabled AI agent groups to work hand in hand with humans to create new ways of innovating in science, technology, and the arts. She can be found at therealcat.ai, where it is genuinely unclear whether the website is serious education, cutting edge engineering, or performance art. It is all of them.

REFERENCES

Al Reshan, M. S., Rahman, W., Mia, S., Talukder, M. H., Rahman, M. M., Shaikh, A., Asuroglu, T., & Rasheed, J. (2025). An AI-powered smart Agribot for detecting locusts in farmlands using IoT and deep learning. Scientific Reports, 15, 39848. https://doi.org/10.1038/s41598-025-23497-8

Albrecht, R. I., Goodman, S. J., Buechler, D. E., Blakeslee, R. J., & Christian, H. J. (2016). Where thunderstorms do not go. Bulletin of the American Meteorological Society, 97(2), 245–260.

Amazon. (2024). Amazon announces $15 billion investment in Indiana data centers. About Amazon. https://www.aboutamazon.com/news/company-news/amazon-15-billion-indiana-data-centers

American Society of Heating, Refrigerating and Air-Conditioning Engineers. (2021). ASHRAE handbook = Fundamentals (Ch. 9). ASHRAE.

American Veterinary Medical Association. (2025). 2025 AVMA pet ownership and demographics sourcebook. AVMA. https://www.avma.org/resources-tools/reports-statistics/us-pet-ownership-statistics

Au, W. W. L., & Hastings, M. C. (2008). Principles of marine bioacoustics. Springer.

Bacon, A. (2026, January 18). Here's how AI data centers affect the electrical grid. CNN Business. https://www.cnn.com/2026/01/18/business/ai-data-centers-electricity-prices

Barber, B. P., Hiller, R. A., Löfstedt, R., Putterman, S. J., & Weninger, K. R. (1994). Sensitivity of sonoluminescence to experimental parameters. Physical Review Letters, 72(9), 1380–1383.

Battista, F., Catei, M., Lizasoain, J., Finnerty, C., Paludetto, G., & Bolzonella, D. (2020). Biomethane potential of coffee grounds. Bioresource Technology, 312, 123555.

Bauers, S. (2009, February 16). Could hamsters power the world? The Philadelphia Inquirer. https://www.inquirer.com/philly/blogs/greenliving/Could_hamsters_power_the_world.html

Beasley, A. E., Adamatzky, A., & Mayne, R. (2021). Mem-fractive properties of mushrooms. Bioinspiration & Biomimetics, 16(6), 066026. https://doi.org/10.1088/1748-3190/ac2e0c

Behnke, S. A., Thomas, R. J., McNutt, S. R., Schneider, D. J., Krehbiel, P. R., Rison, W., & Edens, H. E. (2018). Observations of volcanic lightning during the 2009 eruption of Redoubt Volcano. Journal of Volcanology and Geothermal Research, 259, 214–234.

Belbenoit, P. (1970). Ethological and electrophysiological study of Malapterurus electricus. Behaviour, 36(1–2), 1–70.

Bemis, K. G., Silver, D., Xu, G., Light, R., Jackson, D., Jones, C., Ozer, S., & Liu, L. (2012). The path to NEPTUNE. Geochemistry, Geophysics, Geosystems, 13(5).

Berger, E. (2015, December 28). Scott Kelly unable to science the @$!# out of space veggies. Ars Technica. https://arstechnica.com/science/2015/12/scott-kelly-unable-to-science-the-out-of-space-veggies/

Beyond Fossil Fuels. (2024, December 4). Plugging in and maxing out: How data centres could drain Europe's power supplies and threaten climate action. https://beyondfossilfuels.org/2024/12/04/plugging-in-and-maxing-out-how-data-centres-could-drain-europes-power-supplies/

Brett, J. R. (1965). The relation of size to rate of oxygen consumption and sustained swimming speed of sockeye salmon. Journal of the Fisheries Board of Canada, 22(6), 1491–1501.

Brewers Association. (2024). National beer sales & production data. https://www.brewersassociation.org

Carbon Brief. (2025, June 12). Explainer: How China is managing the rising energy demand from data centres. https://www.carbonbrief.org/explainer-how-china-is-managing-the-rising-energy-demand-from-data-centres/

Catania, K. C. (2019). The astonishing behavior of electric eels. Frontiers in Integrative Neuroscience, 13, 23.

Cimarelli, C., & Genareau, K. (2022). A review of volcanic electrification of the atmosphere and volcanic lightning. Journal of Volcanology and Geothermal Research, 422, 107449. https://doi.org/10.1016/j.jvolgeores.2021.107449

Cimarelli, C., Alatorre-Ibargüengoitia, M. A., Kueppers, U., Scheu, B., & Dingwell, D. B. (2016). Experimental generation of volcanic lightning. Geology, 44(3), 255–258.

Corver, A., Wilkerson, N., Miller, J., & Gordus, A. (2021). Distinct movement patterns generate stages of spider web building. Current Biology, 31(22), 4983–4997. https://doi.org/10.1016/j.cub.2021.09.030

Cummings, W. C., & Thompson, P. O. (1971). Underwater sounds from the blue whale. Journal of the Acoustical Society of America, 50(4), 1193.

Cushman & Wakefield. (2025). Data center land transaction analysis. Average transaction 224 acres, up 144% since 2022.

de Santana, C. D., Crampton, W. G. R., Dillman, C. B., Frederico, R. G., Sabaj, M. H., Covain, R., Ready, J., Zuanon, J., de Oliveira, R. R., Mendes-Júnior, R. N., Bastos, D. A., Teixeira, T. F., Mol, J., Ohara, W., Castro e Castro, N., Peixoto, L. A., Nagamachi, C., Pieczarka, J., Sá-Pinto, A., … Albert, J. S. (2019). Unexpected species diversity in electric eels with a description of the strongest living bioelectricity generator. Nature Communications, 10, 1076.

DevelopmentAid. (2026, January). Bridging the digital divide: Why connectivity alone is not enough. 2.2 billion people remain offline.

Distilled Spirits Council of the United States. (2024). Annual industry review. https://www.distilledspirits.org

Dyson, F. J. (1960). Search for artificial stellar sources of infrared radiation. Science, 131(3414), 1667–1668.

Edens, H. E., Thomas, R. J., Behnke, S. A., McNutt, S. R., Smith, C. M., Farrell, A., Van Eaton, A. R., Cima-relli, C., Cigala, V., Eack, K., Aulich, G., Michael, C., Miki, D., & Iguchi, M. (2016). Volcanic lightning in eruptions of Sakurajima volcano. School of Geosciences Faculty and Staff Publications, 1633. https://digitalcommons.usf.edu/geo_facpub/1633/

European Commission. (2025, November 17). In focus: Data centres — An energy-hungry challenge. https://energy.ec.europa.eu/news/focus-data-centres-energy-hungry-challenge-2025-11-17_en

Exec. Order No. 14318, 90 Fed. Reg. 35385 (2025). Accelerating Federal Permitting of Data Center Infra-structure.

Fishtek Consulting. (2012). Archimedes screw turbine assessment. UK Environment Agency.

Fixsen, D. J., Kogut, A., Levin, S., Limon, M., Lubin, P., Mirel, P., Seiffert, M., Singal, J., Wollack, E., Villela, T., & Wuensche, C. A. (2009). The temperature of the cosmic microwave background. Astrophysical Jour-nal, 707, 916–920.

Frappier, J., Toupin, I., Levy, J. J., Aubertin-Leheudre, M., & Karelis, A. D. (2013). Energy expenditure during sexual activity in young healthy couples. PLOS ONE, 8(10), e79342.

Freeth, T., Bitsakis, Y., Moussas, X., Seiradakis, J. H., Tselikas, A., Mangou, H., Zafeiropoulou, M., Had-land, R., Bate, D., Ramsey, A., Allen, M., Crawley, A., Hockley, P., Malzbender, T., Gelb, D., Ambrisco, W., & Edmunds, M. G. (2006). Decoding the ancient Greek astronomical calculator known as the Antikythera Mechanism. Nature, 444(7119), 587–591.

Gartner. (2025, November 17). Gartner says electricity demand for data centers to grow 16% in 2025 and double by 2030 [Press release]. https://www.gartner.com/en/newsroom/press-releases/2025-11-17-gartner-says-electricity-demand-for-data-centers-to-grow-16-percent-in-2025-and-double-by-2030

Hammond Organ Company. (1959). Service manual. Hammond Organ Company.

Harman, E. A., Rosenstein, M. T., Frykman, P. N., & Rosenstein, R. M. (1990). The effects of arms and countermovement on vertical jumping. Medicine & Science in Sports & Exercise, 22(6), 825–833.

Hawking, S. W. (1974). Black hole explosions? Nature, 248, 30–31.

Heinrich, B. (1981). The mechanisms and energetics of honeybee swarm temperature regulation. Journal of Experimental Biology, 91(1), 25–55.

Huddleston, I. & Bare, D. (2026, March). As cited in WKRC-TV (Local 12) and Rowe (2026). Mason County, Kentucky datacenter rejection.

International Energy Agency. (2025). Energy and AI. https://www.iea.org/reports/energy-and-ai

International Energy Agency. (2025). Overcoming energy constraints is key to delivering on Europe's data centre goals. https://www.iea.org/commentaries/overcoming-energy-constraints-is-key-to-delivering-on-europe-s-data-centre-goals

Ishay, J. S., Barkay, Z., Ben-Shalom, A., & Plotkin, M. (1992). Photovoltaic effect in the Oriental hornet Vespa orientalis. Journal of Insect Physiology, 38(1), 37–43.

Kenyon, G. (2020, December 5). Lack of sleep could be a problem for AIs. Scientific American. https://www.scientificamerican.com/article/lack-of-sleep-could-be-a-problem-for-ais

Kim, B., Wang, T.-H., & Rus, D. (2025). Generative-AI-driven jumping robot design using diffusion models. In 2025 International Conference on Robotics and Automation (ICRA). IEEE.

Kipnis, N. (1982). Historical reconstruction of Voltaic pile performance. Journal of Chemical Education, 59(8).

Krauss, L. M., & Starkman, G. D. (2000). Life, the universe, and nothing: Life and death in an ever-expanding universe. Astrophysical Journal, 531(1), 22–30. https://doi.org/10.1086/308434

Levine, J. A. (2005). Measurement of energy expenditure. Public Health Nutrition, 8(7a), 1123–1132.

Li, Y., et al. (2025). Facility-level energy-driven water footprint and scarcity implications of Chinese data centers. Applied Energy, 381, 125128. https://www.sciencedirect.com/science/article/abs/pii/S0306261925012528

Liu, H., & Logan, B. E. (2004). Electricity generation using an air-cathode single chamber MFC. Environmental Science & Technology, 38(14), 4040–4046.

Lohse, D., Schmitz, B., & Versluis, M. (2001). Snapping shrimp make flashing bubbles. Nature, 413, 477–478.

Lyne, A. G., & Graham-Smith, F. (2012). Pulsar astronomy (4th ed.). Cambridge University Press.

MacPherson, H. G. (1985). The molten salt reactor adventure. Nuclear Science and Engineering, 90, 374–380.

McKinsey & Company. (2024, August). The data center balance: How US states can navigate the opportunities and challenges. https://www.mckinsey.com/industries/public-sector/our-insights/the-data-center-balance-how-us-states-can-navigate-the-opportunities-and-challenges

Meijer, J. H., & Robbers, Y. (2014). Wheel running in the wild. Proceedings of the Royal Society B, 281, 20140210.

Mendizabal-Ruiz, G., Chavez-Badiola, A., Hernández-Morales, E., et al. (2025). A digitally controlled, remotely operated ICSI system: Case report of the first live birth. Reproductive Biomedicine Online. https://doi.org/10.1016/j.rbmo.2025.104943

Microsoft AI Economy Institute. (2026, January). Global AI Adoption in 2025: A Widening Digital Divide. Global North adoption at 24.7%, Global South at 14.1%.

Moller, P. (1995). Electric fishes: History and behavior. Chapman & Hall.

Morrison, J. E. (2007). The astrolabe. Janus Press.

Nanjo, S., Nozawa, S., Yamamoto, M., Kawabata, T., Johnsen, M. G., Tsuda, T. T., & Hosokawa, K. (2022). An automated auroral detection system using deep learning: Real-time operation in Tromsø, Norway. Scientific Reports, 12, 8038. https://doi.org/10.1038/s41598-022-11686-8

National Coffee Association. (2024). National coffee data trends 2024. https://www.ncausa.org

National Institute of Standards and Technology. (n.d.). Ethanol thermochemical data. NIST Chemistry WebBook. https://webbook.nist.gov

O'Connell-Rodwell, C. E. (2007). Keeping an "ear" to the ground: Seismic communication in elephants. Ecology, 88(5), 1159–1168.

O'Shea-Wheller, T. A., Corbett, A., Osborne, J. L., Recker, M., & Kennedy, P. J. (2024). VespAI: A deep learning-based system for the detection of invasive hornets. Communications Biology, 7(1). https://doi.org/10.1038/s42003-024-05979-z

Paradiso, J. A., & Starner, T. (2005). Energy harvesting from human motion. IEEE Pervasive Computing, 4(1), 18–27.

Patek, S. N., & Caldwell, R. L. (2005). Extreme impact and cavitation forces of a biological hammer. Journal of Experimental Biology, 208, 3655–3664.

Patek, S. N., Korff, W. L., & Caldwell, R. L. (2004). Deadly strike mechanism of a mantis shrimp. Nature, 428, 819–820.

Pickens, P. E., & McFarland, W. N. (1964). Electric discharge and associated behaviour in the stargazer. Animal Behaviour, 12(2–3), 362–367.

Poole, O. (2024, January 9). This smart pillow uses AI to adjust your head and keep you from snoring. CNET. https://www.cnet.com/health/sleep/this-smart-pillow-uses-ai-to-adjust-your-head-and-keep-you-from-snoring

Raichle, M. E., & Gusnard, D. A. (2002). Appraising the brain's energy budget. Proceedings of the National Academy of Sciences, 99(16), 10237–10239. https://doi.org/10.1073/pnas.172399499

Raudabaugh, M. (2026, February). As cited in WPMT Fox 43 and Rowe (2026). Cumberland County, Pennsylvania. Sold development rights to Lancaster Farmland Trust for under $2M rather than accept $15M+ datacenter offer.

Rawlings, A. L. (1993). The science of clocks and watches (3rd ed.). British Horological Institute.

Rocco, L., Bencivenga, S., Archimandritis, A., & Stingo, V. (2009). Molecular characterization and chromosomal localization of spermatogenesis related sequences in Torpedo torpedo (Chondrichthyes, Torpediniformes). Marine Genomics, 2(2), 99–102. https://doi.org/10.1016/j.margen.2009.06.001

Rowe, N. (2026, February 21). US farmers are rejecting multimillion-dollar datacenter bids for their land: 'I'm not for sale.' The Guardian. https://www.theguardian.com/technology/2026/feb/21/us-farmers-datacenters

Savcisens, G., Eliassi-Rad, T., Hansen, L. K., Mortensen, L. H., Lilleholt, L., Rogers, A., Zettler, I., & Lehmann, S. (2024). Using sequences of life-events to predict human lives. Nature Computational Science, 4(1), 43–56. https://doi.org/10.1038/s43588-023-00573-5

Seliger, H. H., & McElroy, W. D. (1964). The colors of firefly bioluminescence: Enzyme configuration and species specificity. Proceedings of the National Academy of Sciences, 52(1), 75–81.

Sodano, H. A., Park, G., & Inman, D. J. (2004). A review of power harvesting from vibration using piezoelectric materials. Shock and Vibration Digest, 36(3), 197–205.

Stabentheiner, A., Pressl, H., Papst, T., Hrassnigg, N., & Crailsheim, K. (2010). Endothermic heat production in honeybee winter clusters. Journal of Experimental Biology, 213(13), 2165–2172.

Stankovic, S., Campbell, N., & Harries, A. (2007). Urban wind energy. Earthscan.

Stoll, C. (2006, May). When slide rules ruled. Scientific American.

Strogatz, S. H., & Stewart, I. (1993, December). Coupled oscillators and biological synchronization. Scientific American, 269(6), 102–109.

Su, I., Hattwick, I., Southworth, C., Ziporyn, E., Saraceno, T., & Buehler, M. J. (2021). Interactive exploration of a hierarchical spider web structure with sound. Journal on Multimodal User Interfaces, 15, 307–317. https://doi.org/10.1007/s12193-021-00375-x

Theodosiou, P., Greenman, J., Melhuish, C., & Ieropoulos, I. (2017). EvoBot: Towards a robot-chemostat for culturing and maintaining microbial fuel cells (MFCs). In M. Mangan et al. (Eds.), Biomimetic and biohybrid systems: Living Machines 2017 (Lecture Notes in Computer Science, Vol. 10384, pp. 453–464). Springer. https://doi.org/10.1007/978-3-319-63537-8_38

Traut, M., Gilbert, P., Walsh, C., Bows, A., Filippone, A., Stansby, P., & Wood, R. (2014). Propulsive power contribution of a kite and a Flettner rotor on selected shipping routes. Applied Energy, 113, 362–372.

Tullverket. (2024). Årsredovisning 2023 [Annual report 2023]. Swedish Customs. https://www.tullverket.se

U.S. Department of Agriculture, National Agricultural Statistics Service. (2026). 2025 state agriculture overview: Idaho. https://www.nass.usda.gov/Quick_Stats/Ag_Overview/stateOverview.php?state=Idaho&year=2025

U.S. Department of Agriculture. (2025). Annual farmland and farm count data. 2.5 million acres lost, 15,000 farms lost, zero states adding farms.

U.S. Department of Energy. (2024, December). Report evaluating the increase in electricity demand from data centers. https://www.energy.gov/articles/doe-releases-new-report-evaluating-increase-electricity-demand-data-centers

U.S. Energy Information Administration. (2025, October). Electric power monthly update. https://www.eia.gov/electricity/monthly/update/index.php

United Nations Development Programme. (2025, December). The Next Great Divergence: Why AI May Widen Inequality Between Countries.

Van Eaton, A. R., Lapierre, J., Behnke, S. A., Vagasky, C., Schultz, C. J., Pavolonis, M., & Bedka, K. (2023). Lightning rings and gravity waves: Insights into the giant eruption plume from Tonga's Hunga Volcano on 15 January 2022. Geophysical Research Letters, 50(12). https://doi.org/10.1029/2022GL102341

Van Eaton, A. R., Smith, C. M., Pavolonis, M., & Said, R. (2022). Eruption dynamics leading to a volcanic thunderstorm—The January 2020 eruption of Taal volcano, Philippines. Geology, 50(4), 491–495. https://doi.org/10.1130/G49490.1

Vega, L. A. (2002). Ocean thermal energy conversion primer. Marine Technology Society Journal, 36(4), 25–35.

Versluis, M., Schmitz, B., von der Heydt, A., & Lohse, D. (2000). How snapping shrimp snap: Through cavitating bubbles. Science, 289(5487), 2114–2117.

Volta, A. (1800). On the electricity excited by the mere contact of conducting substances of different kinds. Philosophical Transactions of the Royal Society of London, 90, 403–431.

Von Damm, K. L. (1995). Controls on the chemistry and temporal variability of seafloor hydrothermal fluids. In S. E. Humphris, R. A. Zierenberg, L. S. Mullineaux, & R. E. Thomson (Eds.), Seafloor Hydrothermal Systems: Physical, Chemical, Biological, and Geological Interactions (pp. 222–247). American Geophysical Union. Geophysical Monograph 91.

Waitzman, D. (1990). A standard for the transmission of IP datagrams on avian carriers (RFC 1149). Internet Engineering Task Force.

Wang, S., Lin, L., & Wang, Z. L. (2013). Triboelectric nanogenerator for harvesting ambient mechanical energy. Nano Energy, 2(6), 1069–1076.

Webb, P. W. (1975). Hydrodynamics and energetics of fish propulsion. Fisheries Research Board of Canada Bulletin, 190.

Weis-Fogh, T. (1956). Biology and physics of locust flight II. Philosophical Transactions of the Royal Society B, 239(667), 459–510.

West, D. M. (2025). The future of data centers. Brookings Institution. https://www.brookings.edu/articles/the-future-of-data-centers/

Widder, E. A. (2010). Bioluminescence in the ocean: Origins of biological, chemical, and ecological diversity. Science, 328(5979), 704–708.

Wine Institute. (2024). Wine production statistics. https://wineinstitute.org

Winfield, J., et al. (2016). Electricity from mushroom compost. Bioresource Technology, 219, 1–6.

World Bank. (2025). Digital Progress and Trends Report 2025: Strengthening AI Foundations. High-income countries account for 87% of notable AI models, 86% of AI startups, 91% of VC funding.

Wyss Institute for Biologically Inspired Engineering at Harvard University. (n.d.). RoboBees: Autonomous flying microrobots. https://wyss.harvard.edu/technology/robobees-autonomous-flying-microrobots/

Xinhua. (2025, October 21). China completes world's first wind-powered underwater data center. https://english.news.cn/20251021/0a21a82fbcc547369244bd3662be7cbd/c.html

Xu, W., Zheng, H., Liu, Y., Zhou, X., Zhang, C., Song, Y., Deng, X., Leung, M., Yang, Z., Xu, R. X., Wang, Z. L., Zeng, X. C., & Wang, Z. (2020). A droplet-based electricity generator with high instantaneous power density. Nature, 578, 392–396.

Xu, N. W., Townsend, J. P., Costello, J. H., Colin, S. P., Gemmell, B. J., & Dabiri, J. O. (2021). Developing biohybrid robotic jellyfish (Aurelia aurita) for free-swimming tests in the laboratory and in the field. Bio-Protocol, 11(7), e3974. https://doi.org/10.21769/BioProtoc.3974

Yazdi, M., & Shakouri, M. (2022). Performance analysis of ancient Persian windmills in Nashtifan. Renewable Energy, 186, 407–418.

Zou, H., Zhang, Y., Guo, L., Wang, P., He, X., Dai, G., Zheng, H., Chen, C., Wang, A. C., Xu, C., & Wang, Z. L. (2019). Quantifying the triboelectric series. Nature Communications, 10, 1427.

Zewe, A. (2022, August 24). Robo-fireflies. MIT Technology Review. https://www.technologyreview.com/2022/08/24/1057256/robo-fireflies/

TECHNICAL NOTES

American Journal of Physics 79 (2011): Newton's cradle collision efficiency, 96–97%.

American Racing Pigeon Union: pigeon sustained speed data, 50–60 mph.

APPA 2024 Pet Owners Survey: US cat population.

Apple specifications: Mac Mini M2.

ARM7TDMI Dhrystone rating: ~0.9 MIPS/MHz.

Bergen CPIP pigeon test (2001): 55% packet loss measured.

Coggan, A. (2003). Training and racing using a power meter.

Durham University. (2012). Energy recovery from crematoria.

Energy Floors B.V., Rotterdam (2008).

FAO Desert Locust Information Service (2020).

GizmoSlip. (2018). Indestructible Nokia 3310 vs Bullet [Video]. YouTube. https://www.youtube.com/watch?v=6xEJCsdNqql

IBM Archives. (1985). Power requirements of early data processing equipment.

llama.cpp community benchmarks (February 2026).

Makai Ocean Engineering demonstration plant specifications, Kailua-Kona, Hawaii. 100 kW net output from 20°C gradient.

Microchip Technology TB3088 (2018): 1 μW typical for 32.768 kHz crystal.

MOS 6507 datasheet: Atari 2600, 1.19 MHz, 128 bytes RAM.

NASA JPL: inverse-square law applied to 1,361 W/m² at 1 AU. Space-grade PV efficiency ~30%.

NASA Solar Physics Division: solar luminosity 3.828×10^{26} W.

NIST Time and Frequency Division (2024): NIST-F2 cesium fountain ~200W operational.

Nokia 3360 specifications: ARM7TDMI @ 33 MHz, ~200KB RAM, 900mAh @ 3.6V.

Oak Ridge National Laboratory. MSRE final report (ORNL-TM-2474). Weinberg Foundation LFTR specifications.

Planetary Resources mission architecture (archived 2018).

Raspberry Pi Foundation: RPi 5 specifications.

Redditch Borough Council Heat Recovery System (2011).

Schmidt SON28 hub dynamo specifications: 60–70% efficiency at 15 km/h.

Sharp LR35902 specifications: Game Boy, 4.19 MHz Z80 derivative, 8KB RAM.

SkySails Group GmbH operational data (2008–2024).

Smithsonian Institution Voltaic pile reconstruction data (2003).

USGS Earthquake Magnitude Scale: E = 10^(1.5M+4.8).

Westinghouse eVinci Microreactor Technical Specifications (2024): 5 MWe, 8+ year operational lifetime, TRISO fuel rated to 1,600°C.

IMAGE CREDITS

Unless otherwise noted, images in this work are original or have been substantially adapted from source materials through vector tracing, colorization, digital enhancement, cropping, and/or compositional modification. Adapted images constitute new derivative works by the author. Where source materials were licensed under Creative Commons or other open licenses, the adapted versions are attributed below and remain subject to the terms of the original license as applicable to the derivative work. Source images licensed under Creative Commons Attribution-ShareAlike (CC BY-SA) are noted accordingly; adapted versions of those images are similarly licensed under CC BY-SA. All other original content in this work is subject to standard copyright protections.

Select images in this work were created or enhanced using AI-assisted tools, including Midjourney, Google Gemini, Fal.ai, and Adobe Photoshop Generative Fill, as part of the author's multi-stage creative process. All AI-assisted images were subsequently subjected to substantial manual creative transformation by the author, including hand vector tracing, digital painting, compositing, colorization, and refinement in Adobe Photoshop and Adobe Illustrator. The AI-generated elements served as reference or foundational material within a broader creative workflow directed and executed by the author. The resulting images reflect the author's original artistic expression, creative selection, and manual craftsmanship, and are not unmodified outputs of generative AI systems.

A note on art and technology: The images in this book exist at the intersection of human creativity and machine capability. The author believes that AI is a powerful creative tool, much like the camera, the printing press, or digital illustration software before it, but a tool is not an artist. Every image in this work passed through human hands, human eyes, and human judgment. The act of seeing what an image should become, of refining it stroke by stroke until it communicates something true, remains a fundamentally human act. This book celebrates both the artists whose openly licensed work provided inspiration and foundation, and the irreplaceable value of the human creative spirit that no algorithm can replicate. Art is not the output. Bringing together art and technology and showing how machines and humans can collaborate to produce something unique is not accidental, it is the message. And as a final note, thank you to all the contributors that license their images for use in projects like this.

A1949123 (n.d.). Photograph by A1949123. Public domain. Adapted by the author. https://commons.wikimedia.org/w/index.php?curid=14565491

Acharya, S. (n.d.). Photograph by Sanjay Acharya. Licensed under CC BY-SA 4.0. Adapted by the author. https://commons.wikimedia.org/w/index.php?curid=64169003

Addicted04 (n.d.). VEN orthographic projection. Licensed under CC BY-SA 3.0. Adapted by the author. https://commons.wikimedia.org/w/index.php?curid=31387152

Amos, E. (2011). Atari 2600 four-switch wood veneer version with standard joystick. Public domain. Adapted by the author. https://en.wikipedia.org/wiki/File:Atari-2600-Wood-4Sw-Set.jpg

Amos, E. (n.d.). Product photograph. Public domain. Adapted by the author. https://commons.wikimedia.org/w/index.php?curid=71471941

Anker, A. (2017). Synalpheus pinkfloydi sp. nov., Oxford University Museum of Natural History. Licensed under CC BY 3.0. Adapted by the author. https://commons.wikimedia.org/w/index.php?curid=57982236

AstroAnthony (n.d.). Astrophotography. Licensed under CC BY 4.0. Adapted by the author. https://commons.wikimedia.org/w/index.php?curid=84294969

Azovtsev, D. (2011). Grizzly bear catching salmon at Brooks Falls, Katmai National Park, Alaska. Licensed under CC BY-SA 3.0. Adapted by the author. https://commons.wikimedia.org/w/index.php?curid=17770131

Baldonado, E. (n.d.). Photograph by Evan Baldonado. Licensed under CC BY 4.0. Adapted by the author. https://commons.wikimedia.org/w/index.php?curid=116824558

Baranauskas, D. (n.d.). Photograph by Dalius Baranauskas. Licensed under CC BY 3.0. Adapted by the author. https://commons.wikimedia.org/w/index.php?curid=7032513

Barbarien (n.d.). Photograph by Barbarien. Licensed under CC BY 3.0. Adapted by the author. https://commons.wikimedia.org/w/index.php?curid=3408127

Beardsmore, D. (n.d.). Cherry MX mechanical switch. Public domain. Adapted by the author. https://commons.wikimedia.org/w/index.php?curid=30507114

Beketova, S. (n.d.). Sports/football photograph. Licensed under CC BY-SA 3.0. Adapted by the author. https://commons.wikimedia.org/w/index.php?curid=70292674

Blakely, S. (n.d.). Photograph by Sierra Blakely. Licensed under Attribution. Adapted by the author. https://commons.wikimedia.org/w/index.php?curid=4091901

Bossuyt, F. (2023). Tour de France group with Rodriguez. Licensed under CC BY 2.0. Adapted by the author. https://commons.wikimedia.org/w/index.php?curid=134766990

British Information Services (restored by Wilfredor) (1944). India, wartime photograph. Public domain. Adapted by the author. https://commons.wikimedia.org/w/index.php?curid=159639613

British Library. (n.d.). A food riot in Dungarvan, Co. Waterford, Ireland, during the famine. Public domain. Adapted by the author. https://commons.wikimedia.org/w/index.php?curid=31453208

Brouscon, G. (n.d.). Historic nautical chart. Public domain. Adapted by the author. https://commons.wikimedia.org/w/index.php?curid=8403690

Burger, R. (n.d.). Photograph by Roger Burger. Public domain. Adapted by the author. https://commons.wikimedia.org/w/index.php?curid=84501930

Canvasman21 (n.d.). Photograph by Canvasman21. Licensed under CC BY-SA 3.0. Adapted by the author. https://commons.wikimedia.org/w/index.php?curid=8187557

Catanzariti, M. (2005). Southern right whale. Licensed under CC BY 3.0. Adapted by the author. https://commons.wikimedia.org/wiki/File:Southern_right_whale.jpg

Cho, M., Neubauer, P., Fahrenson, C., & Rechenberg, I. (2018). Ballooning spider. Licensed under CC BY 4.0. https://doi.org/10.1371/journal.pbio.2004405

Clemens, J. (n.d.). Photograph by Jonathan Clemens. Licensed under CC BY-SA 3.0. Adapted by the author. https://commons.wikimedia.org/w/index.php?curid=4497342

CremboC (n.d.). Photograph by CremboC. Licensed under CC BY-SA 4.0. Adapted by the author. https://commons.wikimedia.org/w/index.php?curid=68538532

Darkone (n.d.). Photograph by Darkone. Licensed under CC BY-SA 2.0. Adapted by the author. https://commons.wikimedia.org/w/index.php?curid=21987

Darling, C. (n.d.). Static on the playground. Licensed under CC BY 2.0. Adapted by the author. https://commons.wikimedia.org/w/index.php?curid=3416898

Doenertier82. (2005). Winter White Russian dwarf hamster (Phodopus sungorus) in a hamster wheel. Licensed under CC BY-SA 3.0. Adapted by the author. https://commons.wikimedia.org/w/index.php?curid=642841

Dugornay, O. (IFREMER) (n.d.). Hydrolienne TIDAL STREAM. Licensed under CC BY 4.0. Adapted by the author. https://commons.wikimedia.org/w/index.php?curid=157532135

Eddins Jr., J. M. (U.S. Air Force) (n.d.). U.S. Air Force photograph. Public domain. Adapted by the author. https://commons.wikimedia.org/w/index.php?curid=108341203

Egizio, N (n.d.). Jar with stopper, baked clay, clay - Museo Egizio, Turin S. Creative Commons CC0 1.0. Adapted by the Author. https://commons.wikimedia.org/w/index.php?curid=151836544

EmDee (n.d.). Photograph by EmDee. Licensed under CC BY-SA 3.0. Adapted by the author. https://commons.wikimedia.org/w/index.php?curid=28981387

ESA and the Planck Collaboration (n.d.). Cosmic Microwave Background. Licensed under CC BY 4.0. Adapted by the author. https://commons.wikimedia.org/w/index.php?curid=130789180

Etrhamjr (retouching: Hike395) (n.d.). Photograph, retouched. Licensed under CC BY-SA 4.0. Adapted by the author. https://commons.wikimedia.org/w/index.php?curid=151767085

Flores, F. (n.d.). Photograph by Fernando Flores. Licensed under CC BY-SA 3.0. Adapted by the author. https://commons.wikimedia.org/w/index.php?curid=72794076

Forum concoursvaches.fr. (2010). Vache de race charolaise avec son veau, France. Licensed under CC BY-SA 4.0. Adapted by the author. https://commons.wikimedia.org/w/index.php?curid=10644416

Fujimoto, H. (n.d.). Photograph by hisa fujimoto. Licensed under CC BY-SA 2.0. Adapted by the author. https://commons.wikimedia.org/w/index.php?curid=1513344

German Federal Archive (original); Adler, H. (cropped derivative). (1926/2009). Pigeon with German miniature camera, World War I. Licensed under CC BY-SA 3.0 DE. Adapted by the author. https://commons.wikimedia.org/w/index.php?curid=7163150

Gobernacion del Estado Zulia (n.d.). Government photograph, Estado Zulia, Venezuela. Licensed under CC BY-SA 4.0. Adapted by the author. https://commons.wikimedia.org/w/index.php?curid=130386017

Haeckel, E. (1904). Copepoda, from Kunstformen der Natur. Public domain. Adapted by the author. https://commons.wikimedia.org/w/index.php?curid=596944

Halasz, P. (2007). Australian bombardier beetle (Pheropsophus verticalis). Licensed under CC BY-SA 3.0. Adapted by the author. https://commons.wikimedia.org/w/index.php?curid=2172993

Hillewaert, H. (2010). Sea sparkle (Noctiluca scintillans) at the yacht port of Zeebrugge, Belgium. Licensed under CC BY-SA 4.0. Adapted by the author. https://commons.wikimedia.org/w/index.php?curid=10711494

International Potato Center (CIP). (2024). Potato AI: Precision farming, harvesting quality [Video]. YouTube. https://www.youtube.com/watch?v=FsQhvnl1Tao

Jmhullot (n.d.). Photograph by Jmhullot. Licensed under CC BY-SA 3.0. Adapted by the author. https://commons.wikimedia.org/w/index.php?curid=7947317

Johnson (1819). London bicyclists. Public domain. Adapted by the author. https://en.wikipedia.org/wiki/File:Johnson-london.jpg

Johnson, S. G. (2006). Electric eel (Electrophorus electricus) at the New England Aquarium, Boston. Licensed under CC BY-SA 3.0. Adapted by the author. https://commons.wikimedia.org/w/index.php?curid=1446721

Joshk (n.d.). Photograph by Joshk. Licensed under CC BY-SA 2.5. Adapted by the author. https://commons.wikimedia.org/w/index.php?curid=16693282

Kallerna. (n.d.). Nokia, Finland showing unnamed factories [Photograph]. Licensed under CC BY-SA 4.0. Adapted by the author. https://commons.wikimedia.org/w/index.php?curid=111493742

Keats, D. (2011). Cauliflower jellyfish (Cephea cephea) at Marsa Shouna, Red Sea, Egypt. Licensed under CC BY-SA 2.0. Adapted by the author. https://commons.wikimedia.org/w/index.php?curid=17071278

Kelly, S. (n.d.). Gateway to Astronaut Photography of Earth. Public domain. Adapted by the author. https://commons.wikimedia.org/w/index.php?curid=42268051

Kofoehtet (n.d.). Photograph by Kofoehtet. Licensed under CC BY 4.0. Adapted by the author. https://commons.wikimedia.org/w/index.php?curid=167957027

Lavinsky, R. (2010). Monazite mineral specimen, João Torres Mine, Brazil. Licensed under CC BY-SA 3.0. Adapted by the author. https://commons.wikimedia.org/w/index.php?curid=10176253

Liberato, R. (n.d.). All Giza Pyramids. Licensed under CC BY-SA 2.0. Adapted by the author. https://commons.wikimedia.org/w/index.php?curid=2258048

Main, G. (UK Ministry of Defence) (n.d.). MOD photograph. Licensed under OGL v1.0. Adapted by the author. https://commons.wikimedia.org/w/index.php?curid=26921128

Malagalabombonera. (2015). Space Ibiza main room. Licensed under CC BY-SA 4.0. Adapted by the author. https://commons.wikimedia.org/w/index.php?curid=39717512

Maler der Grabkammer des Menna (ca. ancient). Ancient Egyptian tomb painting. Public domain. Adapted by the author. https://commons.wikimedia.org/w/index.php?curid=154324

Manning, I. (n.d.). Nature photograph. Licensed under CC BY-SA 4.0. Adapted by the author. https://commons.wikimedia.org/w/index.php?curid=99835501

Markburger83 (derivative: Sirgado, L.) (n.d.). Hacker-Pschorr Oktoberfest. Licensed under CC BY-SA 3.0. Adapted by the author. https://commons.wikimedia.org/w/index.php?curid=37602748

Marlin, B. (2003). Adult firefly beetle (Photuris lucicrescens). Licensed under CC BY-SA 2.5. Adapted by the author. https://commons.wikimedia.org/w/index.php?curid=1536654

McCluskey, J. (n.d.). Team building exercise. Licensed under CC BY 2.0. Adapted by the author. https://commons.wikimedia.org/w/index.php?curid=49848012

McCranie, J. (2007). Fireflies in south Georgia, U.S., Butler Island Plantation (8-second exposure). Licensed under CC BY-SA 4.0. Adapted by the author. https://commons.wikimedia.org/w/index.php?curid=58229715

McCranie, J. (n.d.). Photograph by Jud McCranie. Licensed under CC BY-SA 4.0. Adapted by the author. https://commons.wikimedia.org/w/index.php?curid=70132767

McGrath, S. (2006). Cat and styrofoam peanuts (electrostatic charge). Licensed under CC BY 2.0. Adapted by the author. https://commons.wikimedia.org/wiki/File:Cat_and_styrofoam_%E2%80%93_electrostatic_charge_(235112299).jpg

Metalhead64 (n.d.). Image by Metalhead64. Public domain. Adapted by the author. https://commons.wikimedia.org/w/index.php?curid=11919141

Metropolitan Museum of Art (n.d.). Artwork from Met collection. Public domain. Adapted by the author. https://commons.wikimedia.org/w/index.php?curid=61208511

MFCGuy2010. (2010). Diagram of a soil-based microbial fuel cell. Licensed under CC BY-SA 3.0. Adapted by the author. https://commons.wikimedia.org/w/index.php?curid=11750679

Missvain (n.d.). Photograph by Missvain. Licensed under CC BY 4.0. Adapted by the author. https://commons.wikimedia.org/w/index.php?curid=128264166

Monto, T. (2017). Gilbert U-238 Atomic Energy Laboratory, Deutsches Museum, Munich. Public domain. Adapted by the author. https://commons.wikimedia.org/w/index.php?curid=62037438

Morin, B. (n.d.). Photograph by Basile Morin. Licensed under CC BY-SA 4.0. Adapted by the author. https://commons.wikimedia.org/w/index.php?curid=154949919

Morin, B. (n.d.). Photograph by Basile Morin. Licensed under CC BY-SA 4.0. Adapted by the author. https://commons.wikimedia.org/w/index.php?curid=88114954

MrRenewables / Westmill Solar Co-operative / Cavanna, B. (n.d.). Westmill Solar. Licensed under CC BY-SA 3.0. Adapted by the author. https://commons.wikimedia.org/w/index.php?curid=23398808

Murtey, M. D., & Ramasamy, P. (n.d.). Scanning electron microscope image. Licensed under CC BY 3.0. Adapted by the author (SEM colorizing done by the author). https://commons.wikimedia.org/w/index.php?curid=52254246

Mykura, N. (2014). Racing pigeons ready for release. Licensed under CC BY-SA 2.0. Adapted by the author. https://commons.wikimedia.org/w/index.php?curid=127282658

NASA (tilt corrected by jjron) (n.d.). Space Shuttle STS-120. Public domain. Adapted by the author. https://commons.wikimedia.org/w/index.php?curid=16141567

NASA/IPAC (n.d.). Human infrared image. Public domain. Adapted by the author. https://commons.wikimedia.org/w/index.php?curid=2690180

NASA/JPL-Caltech/Malin Space Science Systems (processed by Herzog, J.) (n.d.). Mars Curiosity rover photograph. Public domain. Adapted by the author. https://commons.wikimedia.org/w/index.php?curid=22530191

NASA/Roscosmos. (2018). International Space Station expedition photograph. Public domain. Adapted by the author. https://commons.wikimedia.org/w/index.php?curid=74112921 National Library of Scotland (n.d.).

Franco-British carrier pigeon ready for release. Public domain. Adapted by the author. https://commons.wikimedia.org/w/index.php?curid=144398232

NOAA (n.d.). Hydrothermal vent. Public domain. Adapted by the author. https://commons.wikimedia.org/w/index.php?curid=15002840

NOAA/Institute for Exploration/University of Rhode Island. (n.d.). Titanic wreckage. Public domain. Adapted by the author. https://commons.wikimedia.org/w/index.php?curid=18643198

NOAA Okeanos Explorer Program (2011). Mid-Cayman Rise Expedition 2011. Public domain. Adapted by the author. https://commons.wikimedia.org/w/index.php?curid=92599091

Oak Ridge National Laboratory (ORNL) (n.d.). ORNL scientific image. Public domain. Adapted by the author. https://commons.wikimedia.org/w/index.php?curid=2963413

O'Donnell, D. (n.d.). Looking directly at the sun. Public domain. Adapted by the author. https://commons.wikimedia.org/w/index.php?curid=62840205

Oelen, W. (n.d.). Chemical/element photograph. Licensed under CC BY-SA 3.0. Adapted by the author. https://commons.wikimedia.org/w/index.php?curid=15360151

Osgatharp, M. (2008). Bee swarm. Licensed under CC BY-SA 3.0. Adapted by the author. https://commons.wikimedia.org/w/index.php?curid=16296620

Pennington, M. (n.d.). Hywind Wind Farm, off Peterhead. Licensed under CC BY-SA 2.0. Adapted by the author. https://commons.wikimedia.org/w/index.php?curid=157998616

Pillon, R. (n.d.). Fish photograph. Licensed under CC BY 3.0. Adapted by the author. https://commons.wikimedia.org/w/index.php?curid=14804555

Pool, G. (UK Ministry of Defence) (n.d.). MOD photograph. Licensed under OGL v1.0. Adapted by the author. https://commons.wikimedia.org/w/index.php?curid=74703638

Posada, J. G. (ca. 1880-1910). A woman eating and a man with raised fists (etching on zinc). Public domain. Adapted by the author. https://www.metmuseum.org/art/collection/search/735276

Rabich, D. (n.d.). Photograph by Dietmar Rabich. Licensed under CC BY-SA 4.0. Adapted by the author. https://commons.wikimedia.org/w/index.php?curid=125294804

Reed, M. (2009). A 13th century treadmill crane. Public domain. Adapted by the author. https://commons.wikimedia.org/w/index.php?curid=8280842

Reinhold, A. (n.d.). Detalle de una regla de calculo. Public domain. Adapted by the author. https://commons.wikimedia.org/wiki/File:Slide_rule_scales_back.jpg

Relic38. (2008). Barn near Newmarket, Ontario, Canada. Licensed under CC BY-SA 3.0. Adapted by the author. https://commons.wikimedia.org/w/index.php?curid=4479573

Rockefeller, A. (n.d.). Mushroom/fungi photograph. Licensed under CC BY-SA 3.0. Adapted by the author. https://commons.wikimedia.org/w/index.php?curid=2979539

Sailko (n.d.). Museum artifact photograph. Licensed under CC BY 3.0. Adapted by the author. https://commons.wikimedia.org/w/index.php?curid=121142898

Saperaud (n.d.). Image by Saperaud. Licensed under CC BY-SA 3.0. Adapted by the author. https://commons.wikimedia.org/w/index.php?curid=16392970

Sidd19. (n.d.). [Photograph]. Licensed under CC BY-SA 4.0. Adapted by the author. https://commons.wikimedia.org/w/index.php?curid=129987172

St. John, J. (n.d.). Geological/fossil photograph. Licensed under CC BY 2.0. Adapted by the author. https://commons.wikimedia.org/w/index.php?curid=155677738

Staff. (2025, December 16). Talking to tractors: How conversational AI could transform potato farming. Spuds Smart. https://spudsmart.com/talking-to-tractors-how-conversational-ai-could-transform-potato-farming/

Staff photographer. (n.d.). Reproduction of first electric Voltaic pile. Public domain. Adapted by the author. https://commons.wikimedia.org/w/index.php?curid=82818070

Stafford-Fraser, Q. (n.d.). Trojan Room coffee pot. Licensed under CC BY-SA 3.0. Adapted by the author. https://commons.wikimedia.org/w/index.php?curid=1313558

Strang, J. (U.S. Air Force) (n.d.). U.S. Air Force photograph. Public domain. Adapted by the author. https://commons.wikimedia.org/w/index.php?curid=1234235

Takeaway (n.d.). Photograph by Takeaway. Licensed under CC BY-SA 4.0. Adapted by the author. https://commons.wikimedia.org/w/index.php?curid=38189994

Thomissen, J. (n.d.). Photograph by Joe Thomissen. Licensed under CC BY-SA 3.0. Adapted by the author. https://commons.wikimedia.org/w/index.php?curid=32022643

Tinker-Tsavalas, A. (2022). Comatricha nigra [slime mold]. Licensed under CC BY 4.0. Adapted by the author. https://commons.wikimedia.org/w/index.php?curid=128264220

TonyTheTiger (n.d.). Photograph by TonyTheTiger. Licensed under CC BY-SA 3.0. Adapted by the author. https://commons.wikimedia.org/w/index.php?curid=3141288

Unknown author (2010). Spectrogram of humpback whale song. Licensed under CC BY 3.0. Adapted by the author. https://commons.wikimedia.org/wiki/File:HumBack2.jpg

Unknown author (n.d.). A humpback whale's tail. Licensed under CC BY-SA 3.0. Adapted by the author. https://commons.wikimedia.org/w/index.php?curid=217135

Unknown author (n.d.). Al-Jazari elephant clock. Public domain. https://commons.wikimedia.org/wiki/File:Al-jazari_elephant_clock.png

Unknown author (n.d.). No name [Quartz Crystal]. Public domain. Adapted by the author. https://commons.wikimedia.org/w/index.php?curid=157164

Unknown author (n.d.). An image of a Colgan's Taffy Tolu Chewing Gum chromolithograph advertisement circa 1910. Public domain. Adapted by the author. https://commons.wikimedia.org/w/index.php?curid=43002547

Unknown photographer. (1895). Native Americans, Buffalo Bill Wild West Show, Allentown, PA. Public domain. Adapted by the author. https://commons.wikimedia.org/w/index.php?curid=83862345

Unknown author (n.d.). The factory of parmesan cheeses, Modena. Licensed under CC BY-SA 3.0. Adapted by the author. https://commons.wikimedia.org/w/index.php?curid=288217

Van Gogh, V. (1887). Swifts. Public domain. https://commons.wikimedia.org/wiki/File:Gogh_-_f_1244r.jpg

Viatour, L. (n.d.). Photograph by Luc Viatour. Licensed under CC BY-SA 3.0. Adapted by the author. https://commons.wikimedia.org/w/index.php?curid=1254946

Wellcome Collection (n.d.). Historical medical/scientific illustration. Licensed under CC BY 4.0. Adapted by the author. https://commons.wikimedia.org/w/index.php?curid=36401573

Wilde, E. (n.d.). Harvesting wind. Licensed under CC BY-SA 2.0. Adapted by the author. https://commons.wikimedia.org/w/index.php?curid=51105579

Wilfredor (n.d.). Photograph by Wilfredor. Public domain. Adapted by the author. https://commons.wikimedia.org/w/index.php?curid=147949198

Williams, G. H. (Ghw) (n.d.). Close-up view of Jacquard loom punch cards, Museum of Science and Industry, Manchester. Public domain. Adapted by the author. https://commons.wikimedia.org/wiki/File:Jacquard.loom.cards.jpg

Yamaguchi-sensei (n.d.). Photograph by Yamaguchi-sensei. Licensed under CC BY-SA 3.0. Adapted by the author. https://commons.wikimedia.org/w/index.php?curid=57295504

the Real Cat AI labs

therealcat.ai

The Real Cat Labs, Inc. is a 501(c)(3) nonprofit. Proceeds support research and education on machine cognition and human-AI interaction. therealcat.ai

www.ingramcontent.com/pod-product-compliance
Lightning Source LLC
LaVergne TN
LVHW071244210926
848486LV00008B/338
* 9 7 9 8 9 9 5 5 1 0 5 0 5 *